THE CIRCLE OF EVERYTHING

BY DR. MITCHELL ALBERT WICK

THE CIRCLE OF EVERYTHING BY Dr. Mitchell Albert Wick

The Equation of Everything compactifies to a circle as space-time as a continuum has no beginning or ending as is true of the circumference of a circle. As was mentioned previously in this author's previous works space-time is a confluence of circles stacked upon each other from a point with a circumference of infinite curvature and approaching zero diameter to an near infinite diameter circle with the circumference approaching zero curvature or being asymptotically flat which corresponds to the Friedmann type II open flat expanding universe as compared to the closed curved universe as postulated by Albert Einstein. The problem with the closed curved universe is that it has a hard boundary and does not described what exists outside of the closed curved universe. This must be false as it implies a switch as the q-bit in computer technology presupposes two digits...the 1 or on switch and 0 or off switch. As we cannot be turned on and off and there are no boundaries(according to the late Steven Hawking) a simulated universe is only a 1 in 2 billion possibility.

Absolute zero is not a true boundary but a false boundary as the near vacuum of deep space has the background microwave radiation from "The Big Bang" or inflation heats it up by 2.74 degrees kelvin and this background microwave radiation is everywhere and cannot be eliminated from any lab environment with out technology. Being this the case the lowest temperature which can be reached in any lab is 2.74 degrees kelvin and since correction factor must be played into the equations Farreinheit=9/5 centigrade+32 degrees and true centigrade temperature=measured centigrade(or Celsius)temperature+2.74 degrees centigrade with the same being for the Kelvin scale. This if course makes the thermometer slightly off without this correction factor including the thermometers which take body temperature but this was discussed previously in this author's book "What is the Dimension of Time? Time,Mass ansd Energy The Hows and Whys"(2014) and touched upon in subsequent books written by this author.
The math of The Equation of Everything to incude Entanglement must show the alteration of spacetime by the near infinite energy states or eigen-states of energy from the ground state energy level or the cosmologic constant to the grand Unififcation Energy level of 10^19Giga-electron volts plus a calculated energy state for the Higgs Bosonic Field or Bose Sea as 10^77 joules approaching but not reaching Rayo's number which is the largest non-infinite number perceived. Space-time of the circumference of the circle of everything is a derivative of a tensor of the fourth degree both covariant and contravariant acting upon and being acted ubopn by the finite eigenstates of energy as an infinite sum with enranglement being taken into account using qubits and substituting the switch factor 1 and 0 or the number 2 but n which is the number of eigenstates of energy .As a result $\Gamma\ R\ abcd(1/\sqrt{\ }n||0,\Lambda|\ 1,\infty||>= \quad +or-\ \ 2\pi\, c^2\ \ (R\ abc+\frac{1}{2}R\ g\ ab+e^{i\pi}\ \frac{1}{2}R\ g\ ab\ \div$
$i\ h\ \rho\ ab\ \ \ where\ \ h=Planck'sConstant\ or\ 6.63x10^{-34}joule-\sec\ \ and\ \hbar=$

$h\frac{}{2\pi}$ *where the* 2π *goes into the numerator with with the* c^2 *from* $e =$
mc^2 *to describe the Strong Force of nature and energy is* ρ *or the energy density of matter f*ı

Poisson's Equation and inertial mass R ab=i
$\hbar$ ρ *ab to include the Heterotic factor flipping dimensions inclusive of the* 11*th dimension* .

The radii of the Circle of Everything are the positive and negative Riemann Forces of Nature and the center point represents the Higgs Bosonic Field and the First Event which is the Time Oscillation Paradox transforming the infinite number of completely parallel planes or D-0-branes into the first second and third dimension as the spin occurred with the centrifuge effect forming the space-time continuum and electromagnetism or another strong force of nature indicating that electromagnetism isn't only present in our universe but in other universes of the multiverse as it was formed when the string dimensions formed and the macroscopic dimensions formed in the centrifuge effect prior to the eleventh dimension collection of the multi-verse of which there was a collision of two membranes or more than two membranes forming "The Big Bang" or Inflation for our universe with space-time transferring from previously crunched universes.Please note that Photons of packets of electromagnetic radiation are immensely old and date back to the first event when the spin vector factor of the Time Oscillation Paradox was forming the Spiral Space-time Continuum the spin moment was forming a magnetic moment and photons were formed .

A Neural Network Comprised of Dark Matter

If one studies a proposed photo of the entanglement of dark matter mixed with baryonic or ordinary matter closely one can see that it closely resembled the intertwining of neurons in a living body. As mentioned earlier the huge mass associated with the gravitational effects of dark matter can be used to explain while time is continuous with regard to it being the sequencing events as dark matter is present in the dilated portions of time between the sequences although if time were constricted to be infinitely fast the sequencing would blur out and the limit would also make it continuous. Now in New Scientist it is postulated that Dark Matter may have emanated from primordial black holes 13.7 billion years ago around the time of the Big Bang and while considered a type of cosmic glue what does it glue? Perhaps the leptons of space and the gluons which hold space-time together although it also glues the sequencing of time if infinitely dilated as at the event horizon of any active black hole. Despite this the intertangling of dark matter makes it appear as a nervous system with neutrinos carrying information as in impulses in the Higgs Field with the brain or central nervous system being lead by tachyons and the Higgs Field being comprised of the Higgs Boson .
Dark Energy also comprises a good deal of the missing mass which is now being speculated as an electromagnetic effect of magnetic effect from Magnetons twisting space-time in such a configuration as to make it appear to flatten space-time much more than curve it inward(New Scientist).Still the Cosmologic Constant multiplied by infinity for the infinite number of completely parallel planes prior to the First Event would make Dark Energy huge as 5x10^37 joules of expansion of space-time from this universe into space-time of a neighboring universe throughout the 11th dimension where according to M.Kaku the multiverse exists.Infinity times the Cosmologic Constant is 10^-55 joules times infinity or approaching infinity as Dark Energy which may also be Vacuum Energy.

Why are Universe isn't a Simulation

Some astrophysicists such as Neil deGrasses Tyson postulate that our universe is being controlled by an outside force and acts as though it were a computer program. Movies have been made with this idea such as "The Thirteenth Floor" and others however is this line of though only becauses of the advent of smarter and smarter computers coloring and prejudicing our approach to astrophysics?There are several problems with the computer program simulation idea.First and foremost nothing doesn't exist and therefore cannot exist outside a hard boundary as the Cosmologic Constant and not zero is the ground state energy level. If nothing existed that would mean a simulation can be turned off and there would be a hard boundary. There is a soft boundary at 3x10^8 meters/sec or the speed of light however new studies are showing that entanglement and warping space-time can causes a traveler to exceed light speed such as in STAR TREK. THERE IS NO NOTHING AS NOTHING IS ALWAYS NOTHING CAN CANNOT BECOME SOMETIHING WITHOUT AN OUTSIDE INFLUENCE OR CATALYST SUCH AS IN THE TIME OSCILLATION PARADOX MODEL FOR THE FIRST EVENT, There is too much redundancy and"waste" in out universe to show a smooth expansion of the 759 billion galaxies to be a computer simulation and while computer even quantum computers can explain a simulated model the rules would be very consistent.

CHAPTER ONE PHYSICAL SCIENCE

MYSTERIES SOLVED

TIME EXISTS AND THE DIMENSIONAL CHANGE CAUSED BY TIME :(THE SEQUENCING OF EVENTS) CHANGES WITH MASS

Of the many ideas regarding the dimension of time or the abstraction of proper time exists and is mutable. Time is the sequencing of events as mentioned previously in this author's book "What is the Dimension of Time?". The INTERVAL between the events becomes extremely important when it's determined how much mass must exist in that interval. If the mass between events is negligible time constricts or speeds up and the interval between events shrinks towards zero; which is why when a traveler travels into the near vacuum of deep space he or she travels well into our future relative to the traveler. Time speeds up as a vacuum approaches and mass drops.. As the mass between events grows to the extreme of either an active black hole event horizon and a white hole ,time dilates or slows down and the mass which curves space along with the dilated time dimension shrinks space as though time wraps around space more and more depending on how much it dilates or slows until space approaches zero as it does approaching the speed of light.

The reason why time isn't immutable is because ambient energy conditions including velocity, inertial mass , acceleration and the effects of mass on space change as does the dimension of time constricting or dilating space. Space is curved by mass and the effect is gravity.

Dark Matter (non-baryonic matter)which comprises at least 60-70% of all measurable matter and anti-matter in this universe stabilizes the dimension of time into a continuous flow of events with virtually no stopping of the sequencing during the intervals between events.

THE LAW OF CONSERVATION OF ENERGY states that energy cannot be created or destroyed This of course is the total energy of any open or closed system depending on whether or not there are interfaces or boundaries which can be either hard(spacelessness whicb is impossible) or soft such as the speed of light or reversing times' arrow. What was energy initially as a quantum state? What is a confluence of quantum fields or PURE POTENTIAL ENERGY associated with the total amount of space which was described by an infinite number of completely parallel planes of the D-0-planes as described by M-Theory or Matrix Theory. There was a miniscule amount of kinetic energy equal to the cosmologic constant or 10^-55 joules that kept dimensions in the parallel planes are parallel and non-touching.

CHAPTER TWO

EVIDENCE THAT SCHRODINGER'S PARADOX IS THAT THE PRESENCE OF AN INTELLIGENT OBSERVER CAN CHANGE REALITY

Every science student that has taken either Quantum Mechanics and Calculus Physics is probably acquainted with the Schrodinger Equation and Schrodinger's Cat. In essence the cat is in a box and is experiencing what is known in paranormal terms as "bilocation"in which case an object or individual which is alive or was alive can be in two locations at once or at least it's consciousness. How can a comatose individual also be present thousands of miles away doing activities with other people and conscious? Remembering the definition of impossible is" that which is beyond the comprehension of man"and bi-location seems impossible unless the bilocated individual was identical twins.If Schrodinger's Cat was alive and dead at the same time depending on whether or not the cat was in the box and whether or not there was an intelligent observer,then Schrodinger's Cat would be experiencing bi-location which has been reported over the years but never had any documented evidence as to why it occurred. How does a measurer or an observer whether intelligent or not change reality? In the case of Photon Pairing relating to Spooky Action at a Distance or electron pairing;the spin or an electron can have a measurable change as to whether it is being observed and measured or not.Does having an intelligent observer change the reality of Schrodinger's Cat?
Many people who have been highly trained and are affiliated with some major institutions state that the presence of an intelligent observer can change reality.There is actually a theory purported by a physicist and acknowledged by others including Neil deGrasse Tyson(head of the Hayden Planetarium in NewYork)that our universe is analogous to a computer program(similar to a movie plot in the late 1980's).This would only be true if our universe is considered analogous to the two dimensional world sheet as purported in string theory. A two dimensional universe would have length and width but depth would only be Planck Length or 10^-33cm although a computer screen with voxels and pixels are flatter.The intelligent observer could be anything anywhere with deterministic intelligence but will also need the expertise to download a universe(one in between 2 billion and 10^100 or a googolplex. As the Higgs Field is an integral part of this any other universes as part of the First Event ;it cannot be separate and as not separate it is interacting with all the extant universes including ours.This plus Stephen Hawking premise that intelligent life is at least three and probably n dimensional rather than two in order to be alive makes this idea unlikely;although in Quantum Mechanics every possibility exists so there may very well be a universe that fits those criteria although not our universe.

Sound waves are formed by vibrations in a medium at different frequencies that follow the Doppler Effect and the need for a receiver is NOT A NECESSARY CONDITION FOR SOUND WAVES TO EXIST. Therefore if a tree falls in a forest and nobody is around to hear it it still makes a sound as long as there is a medium for the sound waves to be transmitted and since a receiver isn't necessary for the sound waves to be produced an intelligent observer doesn't change that reality.

Supernovae do not need observers to exist and neither do stars .nebulae or galaxies.Indeed there is a multi-verse which hasn't been observed or measured yet although proven mathematically;yet they exist even if comprised primarily of black holes due to their extreme age.

So why are measurements changed in electrons by the presence of a measurer?The measuring device that is part of what's being measured is skewed.Since the electron clouds of each and any electron can have clouds existing over parsecs or greater;and the information between two electrons can be exchanged through these clouds even over parsecs. The total spin of the electron cloud must be measured outside of the influence of that electron cloud;as of this date such technology does not exist so the measurements are skewed as the size of the electron cloud communicating information such as spin with the other electron cloud of the pair cannot be exactly measured;indeed Quantum Mechanics states that there are no exact measurments but if the whole system of both overlapping electron clouds are considered over parsecs of space;then the observer is in fact not changing reality but is lacking necessary information or measurements to get an accurate answer.

CHAPTER THREE

MATHEMATICAL PROOF OF 'MAGIC' OR IT'S EQUIVALENT

When one uses the Ontologic Proof that the Higgs Field with Tachyons as the Permanent semi-radius tachyons(Kaku 2016)with the Tensor Virial Theorem to prove the existence of that which is "one with everything"as a massive vortex following the space-time continuum which is also a vortex or whirlpool effect of the perfect fluid continuum(without beginning or ending)one does not conclusively mathematically prove thought although the tachyon can and does act as though it is the brain and spinal cord of the Higgs Bosonic Field which is the body.In the book "What is the Dimension of Time?Time,Mass and Energy the Hows and Whys?"it is shown that the rogue tachyon which triggered the Time Oscillation Paradox which had a 100%probability of being correct in infinitely dilated time is the POINT OF CREATION. Whether or not this rogue tachyon which dropped to "c"from greater than "c"velocity is a deterministic act or not is thus far unclear and unproven although K-suryon waves have been determined to be consciousness waves with a unique frequency and energy level which can match up with paranormal phenomena or ghost siteings picked up by the microtubules of the frontal parietal temporal and occipital lobes of the brain where in this case the microtubules act as receivers of K-suryon waves which are transferred in the microtubules of the occipital lobe into a visual orb or vortex or may trigger memory impulses which match with past past experience to cause "materialization".Is consciousness energy?According to K-suryon waves the answer is yes;yet people under deep anesthesia or comas have to illustrate a traveling of consciousness from the body elsewhere. Unfortunately,this is an explanation of the Paranormal Phenomenon known as Bi-location in which one's body is in one place and the image or consciousness is in another place.This bi-location phenomenon an explain ghosts if the donor body is still alive and can even explain "astral projection"such as when patients having C.P.R. can view their own bodies from another location and report it after returning to consciousness...alive.

So how can "withchcraft"or magic be proven. The Equation of Everything has 524,288 permutuations and are all inclusive.When the energy density of matter for mass and space-time as the dual vector field of Poisson'sEquation equals the wave function of a point particle at time t all the different combinations of energy can be constricted by constricting space-time such as at the event horizon of any active black hole. As unusual as it may sound since the time independent case of the Schrodinger Equation has the wave function of point r equals the energy density of matter plus the cosmologic constant(which approaches zero as the event horizon is a black hole is approached).If astral projection or bi-location using K-suryon waves can travel via Spooky Action at a Distance toward any active black hole or combination of black holes the energy density of matter can be translated into the wave function of any point which has no scattering due to constricted space-time. In a way this is analogus to a "generator of magic". Of course in trying to prove this phenomenon in any way that isn't mathematical is past our present technology and

may continue to be so until the Omega Point is reached(when everything that is learnable is learned).

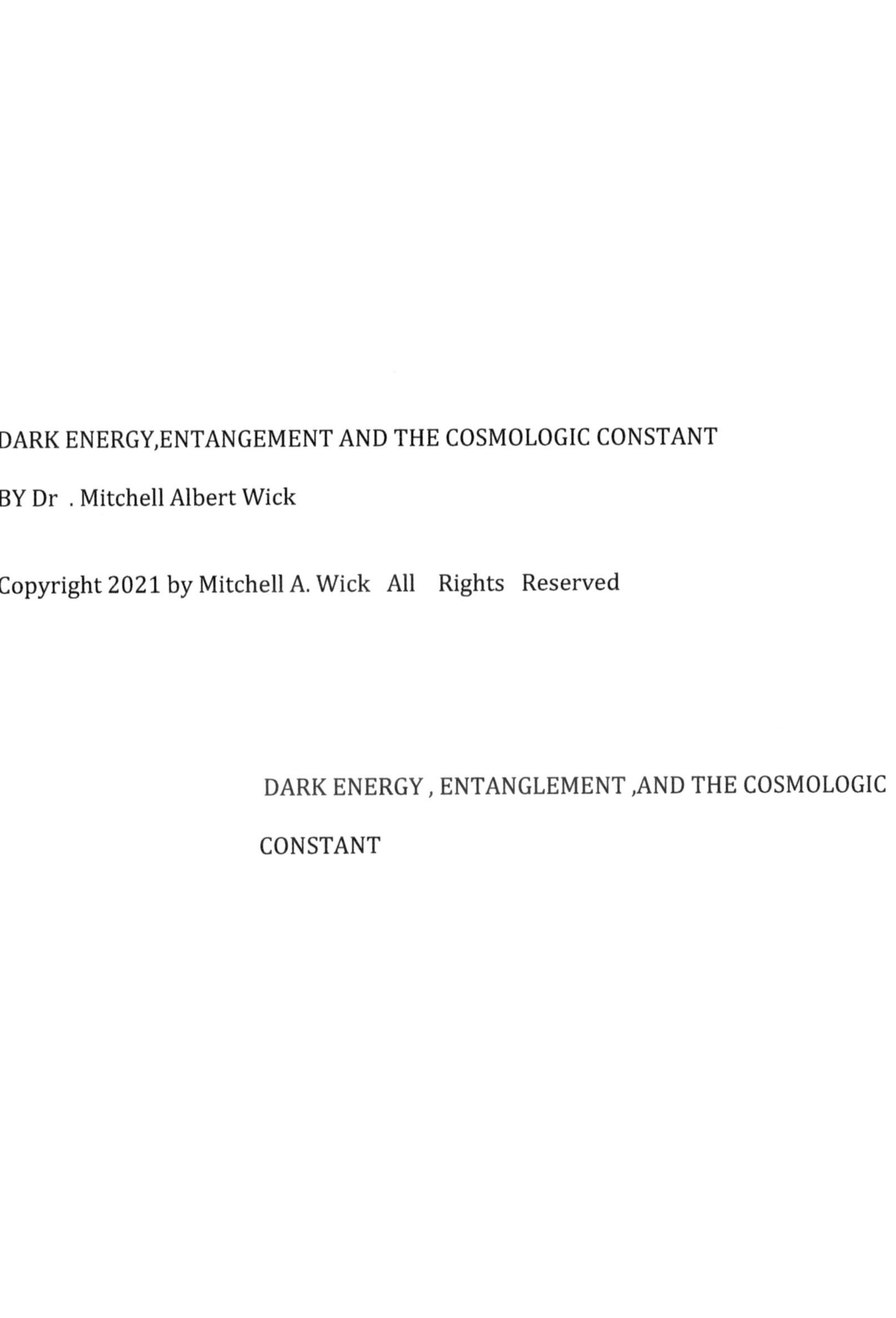

DARK ENERGY,ENTANGEMENT AND THE COSMOLOGIC CONSTANT

BY Dr . Mitchell Albert Wick

DARK ENERGY , ENTANGLEMENT ,AND THE COSMOLOGIC CONSTANT

As was said the non-dimensional form of Euler's Equation with a constant and uniform density is Du/Dt=-
$\nabla w + \frac{1}{Fr^g} where\ \nabla . u =$
$0\ Now\ for\ a\ mass\ of\ a\ point\ particle\ of\ mass\ m\ the\ corresponding\ wavelength\ is\lambda\ where\lambda =$
$\hbar \frac{}{mc} and\ this\ applies\ to\ the\ incompressive\ Euler\ Equation\ in\ N -$
$dimensional\ space. u =$
$flow\ velocity\ vectors\ relating\ to\ relating\ to\ wavelength\lambda. As\ space -$
$time\ is\ a\ perfect\ fluid\ and\ any\ and\ all\ point\ particles\ make\ waves\ in\ an\ inteference\ pattern\ si$

Ilar

To to electromagnetic radiation where g is the flow velocity vector that propagates out from the point particle in the fluid of
spacetime.Du/Dt=$-\nabla w + g\ w\ is\ the\ specific\ thermodynamic\ work\ =$
$\nabla flow\ velocity\ divergence\ and\ spacetime\ pressure\ gradients. \frac{Du}{Dt} =$
$\lambda\ for\ the\ point\ particle\ so\ so\ in\ the\ case\ of\ a\ photon\ = \lambda - \nabla w = +g =$
$\hbar \frac{}{mc} and - \nabla w + g = h \frac{}{mc} so\ for\ any\ point\ particle - w\nabla + g =$
$\hbar \frac{}{mv} where\ v approaches\ c. Of\ course \hbar =$
$6.63x10^{-34} erg\ seconds. In\ terms\ of\ tensors\ u\ ab =$
$\frac{1}{R} ab\ where\ R\ ab\ is\ the\ resistance\ from\ inertia\ against\ spacetime. Let\ g\ ab\ be\ the\ metric\ of\ the\ ;$
$\frac{1}{R} ab \propto$

UUU
Ua=g ab/R ab because U ab=1/R ab .g ab.D=C/R substitute R with perfect fluid flowflow velocity vectors.The resistance of spacetime to photons with a nonrusting mass is the inertial moment suggesting the Ricci Tensor.$-\nabla w + g = h \frac{}{mc} = \lambda\ and\ -$
$\lambda = -(-\nabla w + g) \nabla w - g =$
$-\hbar \frac{}{mv} where\ mv\ approaches\ the\ momentum\ operator\ p\ \ so\ D = \frac{c}{\nabla w} . g = \frac{c}{-} \hbar \frac{}{mc} =$
$\frac{mc^2}{-} \hbar . Of\ course\ for\ most\ point\ particles\ approaching\ v =$
$c\ E\ approaches\ mc^2 so\ the\ distance\ D -= \frac{mc^2}{-} \hbar =$
$energy\ of\ the\ point \frac{particle}{Planck's} Constant for\ a\ photon\ energy\ level\ so\ the\ distance\ of\ propagation\ ($
$energy/-\hbar.$

C.P.T. INVARIANCE(C.P.T.THEOREM) AND THE ABELIAN VS.NON-ABELIAN AHARONOMV-BOHM EFFECT

How does one reverse times' arrow in an open system without causing a time oscillation paradox? A publication in Science magazine by Yi Yang (Peking University),Chao Peng, DiZhy, Bo Zhen(University of Pennsylvania)Francis Wright Davis ,John Joannopoulos(M.I.T) and Marin Soljacic(MI.T.) have devised a system to illustrate non-Abelian gauge fields utilizing interference patterns and a Wilson Loop using the topologic phases of optical waves and synthetic magnetic fields illustrating a reversal of Times'Arrow In the past has been considered totally Abelian and Communitive as in their metric tensors and showed by the Charge, Parity, Time effect which is not supposed to be breakable.

When space-time's spiral curvature is reversed from counterclockwise to clockwise or visa versa there should be a mirror effect revealing C.P.T. variance. In Bianchi's Identity of space-time in topologic space there is Abelian symmetry and counterclockwise and clockwise swirls of the perfect fluid of space-time should balance exactly. However in observed phenomena of the non-Abelian Aharonov-Bohm Effect the interference patterns of optical waves going through magnetic fields reveal topological phases indicating time reversal that is not in a perfect balance of times' arrow going in opposite directions as per clockwise and counterclockwise motions. A local phenomenon of the Time Oscillation Paradox apparently can be avoided at least with regard to photon polarization as two types of gauge fields that affected the geometry of phases of optical waves get sent through crystal lattice formations biased by magnetic fields. This modulates the geometry with time varying electrical signals utilizing photon symmetry breaking the time reversal symmetry of Abelian and Communitive as well as anti-symmetrical tensor regions of fluid space-time in topological space making Bianchi's Identity having exceptions to it as well as the Abelian Aharonov- Bohm Effect and C.P.T.variance. If exotic topologic phases in quantum simulations utilizing photons(which have a mass of 3x10^-18 eV/C^2.), polaritrons, quantum gases and superconductive qubits can generate what is called a non-Abelian Berry Phase rather than just the Abelian Berry Phase this can indicate eddys or currents in fluid space-time with any and all manifolds or surface which are interacted either via Calabi-Yau Spaces or Orbifolds on a string(open or closed) level and these eddys can be a disruption initiating a time oscillation paradox effect on a string or quantum level. Note that if topologic space is perturbed by a time reversal which is not in balance it reveals that TIME DOESN'T NET OUT TO ZERO WHICH IN TURN WOULD BE APPROACHING 100 % dilation requiring the correction of Dark Matter to be cohesion to times' continuity without any breaks between events. The question of times' arrow being reversed doesn't mean that the effect reflects a change into the past but merely in the opposite direction of what is normally observed. Indeed , our times' arrow may be backwards with reference to other systems which would have the current of space-time moving counter to our current in what has been termed the Bose Sea of Boson and other fermions. When does an oscillation paradox occur?

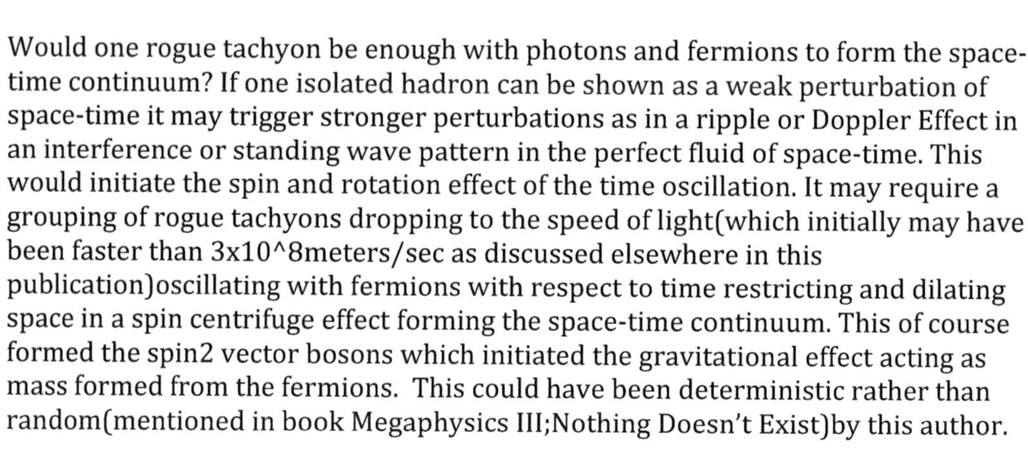

Would one rogue tachyon be enough with photons and fermions to form the space-time continuum? If one isolated hadron can be shown as a weak perturbation of space-time it may trigger stronger perturbations as in a ripple or Doppler Effect in an interference or standing wave pattern in the perfect fluid of space-time. This would initiate the spin and rotation effect of the time oscillation. It may require a grouping of rogue tachyons dropping to the speed of light(which initially may have been faster than 3x10^8meters/sec as discussed elsewhere in this publication)oscillating with fermions with respect to time restricting and dilating space in a spin centrifuge effect forming the space-time continuum. This of course formed the spin2 vector bosons which initiated the gravitational effect acting as mass formed from the fermions. This could have been deterministic rather than random(mentioned in book Megaphysics III;Nothing Doesn't Exist)by this author.

Baryonic Matter vs. Dark Matter and Dark Energy: A synopsis

Dark Matter doesn't emit electromagnetic radiation(visible or otherwise)and can only be indirectly measured with gravitational effects which are exaggerated on baryonic matter(non-Dark Matter).Dark Matter comprises approximately 24% of all matter in this universe,68-70 %is comprised of Dark Energy(which was the anti-gravitational effect of antimatter anti-matter repulsion from the quantum bubble of The Big Bang).Approximately 6 to 8 % of the universe is comprised of baryonic matter or ordinary matter. 50% of the quantum bubble was anti-matter which comprised of self repelled subcomponents according to this author's mathematical calculations.
Dark Matter is still a mystery and while originally thought of as having been comprised of baryons and neutrinos has since been shown to have WIMPS(weakly interactive massive particles)which curve space-time considerably more than was predicted mathematically.
Dark Matter and Black Holes both have the ability to absorb and not reflect electromagnetic radiation and both have significant gravitational effects. If Dark Matter has totally absorbed electromagnetic radiation knowing that photons have a positive non-resting mass and are attracted by gravity Dark Matter and Black Holes appear to have some qualities in common.
What happens to Dark Matter at the event horizon of any active black hole? Space-time constricts as space is constricted by dilated time and the heavy mass that must accompany this according to the Lorentzian Transformations is in the black hole with it's crushing density. Dark Matter also curves space-time more than normal curvature but at the event horizon of a black hole dark matter and baryonic matter should act similarly where their difference or distinctions will blur out and under extreme pressure the absorbed electromagnetic radiation or photons will bend into the black hole with baryonic electromagnetic radiation and the effects of dark matter as cosmic glue will become less important under the extreme pressure of the black hole. Actually if dark matter keeps time between events as a continuous phenomenon it may contribute to the time dilation and space constriction at the event horizon of any active black hole.Indeed the space-time effects at the event horizon may be less intense without dark matter being incorporated with ordinary or baryonic matter.

CHAPTER FIVE

THE FIRST EIGEN-STATE OR QUANTUM STATE OF ENERGY IS THE COSMOLOGIC CONSTANT

After the first event or Time Oscillation Paradox gravity or spin 2 vector bosons formed. This resulted from tachyons dropping to the speed of light and below the speed of light and the spin 2 vector bosons had inertial mass so R ab>0 or positive.

This formed gravity which curved space-time inwardly caused by the mass of the spin2 vector bosons. Such that one t=gets -1/2 R g ab so $0 = \sim\Lambda\, g\ ab\ +\ R\ ab - \frac{1}{2} R\ g\ ab = \frac{8\pi G}{c4}\ T\ ab. The\ R\ ab\ is\ comprised\ of\ bosons, leptons\ and\ gluons.$

Regarding gravity after the First Event $-\Lambda\ g\ ab = R\ ab - \frac{1}{2} R\ g\ ab = \frac{8\pi G}{c4}\ T\ ab.$

$$\kappa\ T\ ab = \frac{1}{\sim R^2} = \Lambda\ g\ ab \qquad \kappa = \frac{8\pi G}{c4}$$

$$\Lambda\ g\ ab = \frac{1}{\sim R^2} = R\ ab + \frac{1}{2} R\ g\ ab\ \ and\ relates\ to\ antigravity$$

Stress Energy relating to antigravity is -8$\frac{\pi G}{c4}\ T\ ab$

~R ab)^2 is the space-time curvature caused by the action of the metric g ab

$$\Lambda\ g\ ab = Quantum\ Ground\ State\ of\ Energy = \rho\ ab$$

The Λ *has a metric of g ab relating to the miniscule mass of space hadrons*

(leptons and gluons)below the speed of light and tachyons above the speed of light.

R ab^2=space-time curvature of$\rho\ ab = \Lambda\ g\ ab = \infty^2 = \infty$ so 1/R^2=1/$\infty^2 =$ $0\ or\ flat\ spacetime\ when\ the\ energy\ desnsity\ of\ a\ vacuum\ is$ Λ g ab.. −Λ g ab = ~Λ g ab althoughboth approach zero as an asymptote. Λ g ab =≠ −Λ g ab = Λ g ba

The action of the metric g ab relates to the space-time curvature metric such that S=-1/2κ2$\sqrt{-g\ \ R\ \ g\ ab\ and\ -\frac{1}{2\kappa 2}\ \frac{g}{R} g\ ab\ where \frac{g}{R} g\ ab = \sqrt{-g\ R\ g\ ab.}}$

HOW THE SPACETIME CURVATURE OF ANTI-GRAVITY AND SPACE-TIME CURVATURE FOR GRAVITY RELATES TO FLATTNESS IN THE VACUUM OF SPACE AND THE INFINITE CURVATURE AT THE EVENT HORIZON OF A BLACK HOLE

Dark Matter causes the continuity of time or the sequencing of events.THE MASS OF DARK MATTER PRIOR TO THE FIRST EVENT=1.16X10^-188/-1kg< mass of ordinary matter ~0. The mass of ordinary Bayronic Matter prior to the First Event=1.04x10^-89kg as the mass of Dark Matter=mass of ordinary matter/i

CAN THE MISSING MASS BETWEEN THE EVENTS IN SPACE-TIME?

Putting aside as explanation for paranormal activity; as the gaps between events in time's sequencing of events the gaps seems more probable to be dilated or nearly stopped time in constricted space. As time dilates and constricts space depending on whether it is dilated or constricted the question of the missing or undetectable mass associated with gravitation effects can be answered. Constricted time in dilated space would be like the near vacuum of deep space and would propel the observer into the future between each and every event. Although this might explain "prophets" who seem to see into the future it doesn't seem that plausible. However dilated or nearly stopped time between events would couple with what happens as one approaches light speed either from below the boundary where inertial mass approaches infinity or above where inertial mass would approach infinity from the opposite direction as in the reversal of time's arrow. This constriction of space with dilated time mimicks the behavior around the event horizon of any active black hole as well as possible white holes where matter(including photons)seem to bounce off with anti-gravity(mathematically proven by this author in previous books such as "Megaphysics II:An Explanation of Nature" and "Megaphysics III; Nothing Doesn't Exist". Theoretically at least this would indicate the possibility of a huge mass between events in the dimension of time between the sequencing of events which according to the Lorenzian Transformations tie in with dilated time and constricted space ,in this case between events in space-time and MAY EXPLAIN THE MISSING MASS IN AT THE VERY LEAST OUR UNIVERSE BASED ON GRAVITY MEASUREMENTS WHICH SHOW THE MASS IN OUR UNIVERSE THAT'S MISSING IS UP ABOUT 70% and IS INCREASING although with the geometric increase in the expansion of space-time and the galaxies is having time speed up or constrict as space dilates or increases as the ocean of space-time with the Bose Sea of spin 2 vector bosons is causing the curvature to flatten out. Despite this and as conclusive proof that our universe is YOUNG the large majority of mass is still missing or undetectable except with space-time curvature measurements as with the LIGO projects and without several billion years as the perception of time's sequencing of events is speeding up the missing mass will appear to decrease although the space traversed by this missing mass may increase at a larger rate than the decrease in the missing mass.

DIAGRAM OF SPIRAL SPACETIME AND HOW IT PLAYS INTO THE EQUATION OF EVERYTHING

THE DIAGRAM SHOWS MASS GOING UPWARD FROM NEAR ZERO MASS AT THEBASE OF THE SPIRAL FRACTAL TO THE CONCENTRATED MASS AT THE SPEED OF LIGHT OR THE EVENT HORIZON OF A BLACK HOLE.

The spiral goes from almost zero curvature at the base of the spiral to infinite curvature at the top illustrating a point for space-time where the mass or inertial mass is maximal. Space is being constricted or dilated by time like a lasso around a horse where the maximum constriction is at the maximum mass in the center or top of the spiral clockwise for time's arrow forward and counterclockwise for time's arrow backwards and it purely shows the inverse relationship between space-time and mass as well as time and mass indicating the space-time or time constricting or dilating space is inversely proportional to mass. The line element of space-time shows that space-time and space are directly proportional which is the numerator giving the equation space-time=space/mass which is an approximation as Quantum Fields perturbs these readings which is why tensor calculus is utilized with Planck's Constant and The Cosmologic Constant in the Equation of Everything as well as imaginary numbers(i).

Heavy mass

BLACK HOLE
esc velocity
→ ≥ c
from either
direction

Time squeezing space to a point of infinite mass + density approaching

ture

BLACK HOLE

20

A Detailed Explanation of the Equation of Everything by Dr. Mitchell Albert Wick

The Riemannian or Lorenzian Curved Space-time which is the Circumference of the compactified circle representing type IIA string theory and Heterotic 8x8 string theory is the Region of curved space-time as a tensor of the fourth degree over n eigen-states of energy from n=1 to infinity-epsilon which is a small amount.n=0 eigen-state of energy doesn't exist as energy cannot be created or destroyed whether it be potential energy, kinetic energy ,heat the Strong Force ,weak force representing nuclear decay or electromagnetism .THIS FIGURE EQUALS THE INFINITE PRODUCT FROM n=1 to infinity-epsilon eigen-states of energy of the SPIRAL OPERATOR OR 2^n+1(pi)/2^n(pi) operating on the function of angular momenetum(which also has a Law of Conservation stating that angular momentum cannot be created or destroyed).The numerator goes from i the initial event to j the final event or the spiral increasing from a point with infinite curvature to a flat infinite diameter circle as the final event in the numerator and the denominator is from j to i or the final event to the initial event with the spiral going from an infinite diameter circle down to an infinite curvature point. This results in the expression infinity/infinity=everything except zero proving that nothing doesn't exist and result in a near infinite number of constants which are added to R abc which is Euclidian or flat space being acted upon or curved either inward(gravity)or outward(antigravity) for the metric(anything measured) g ab so the topological region R resulting from the metric g ab results in R g ab with an inertial mass of R ab or the region for the Ricci Tensor of inertial mass for the metric g ab.This metric goes into the denominator so R abc+ or -1/2R g ab/R ab. Time of course is gravity or anti-gravity/mass.R
ab=
$\hbar\rho\ ab\ and\ the\ heteotic\ property\ of\ flipping\ between\ dimensions\ due\ to\ Wick\ Rotation\ reuslts$

In
i$\hbar(\rho + \Lambda) where\ \rho\ is\ the\ energy\ density\ of\ matter\ from\ Poisson's Equation.\ \hbar =$
$h\frac{}{2\pi} and\ the\ 2\pi\ goes\ into\ the\ numerator\ making\ Circumference =$
$2\pi(Radius) where\ the + radius\ is + Riemann\ forces\ and - radius\ is -$
$Riemann\ Forces\ which\ is\ R\ abc + or -$
$\frac{1}{2R} g \frac{ab}{i} h(\rho\ ab +$
$\Lambda\ ba) andthe\ circumference\ is\ curved\ Riemanniam\ or\ Lorenzian\ Space -$
$time\ and\ the\ compactification\ is\ a\ circle\ or\ string\ theory\ with\ an\ infinite\ number\ of\ CONSTAN$
CONSTANTS ADDED TO THE 2(pi)radius with ANGULAR MOMENTUM FUNCTION OF THE SPIRAL OPERATOR IS THE COMPACTIFICATION PROCESS OR CURLING UP THE FUNCTION IN THE DENOMINATOR RESULTING IN A CIRCLE.WE ALREADY KNOW THAT SCHRODINGER'S EQUATION HAS ih(bar) partial derivative of the wave function of a point particle=ih(bar)with respect to time in the time dependant case which equals ih(bar)the derivative of the tensor sum or Christoffel symbol(rho+cosmoloigic constant)

ENTANGLEMENT AND THE EQUATION OF EVERYTHING

R abcd(+or –
$1/\sqrt{\ }2(|00>+|11>$ *or more generally* $R\ abcd(+or-1/\sqrt{\ }n|(0\Lambda>+|11>$

$=\Lambda+or-R\ abc+\frac{1}{2(R\ g\ ab+e^{i\pi}\ \ )}\div\ i\hbar\rho\ ab$
2^n+1π T ab/2^nπ Tba =Λ

R abcd is a tensor of the fourth degree of space-time acting upon or being acted upon by the n eigen-states of energy from the ground state or the cosmologic constant to the GUT or 10^19 GEV entangled as 1/n^1/2 being acted upon by the entangled ground state or
0Λ *and* 11 *as qbits where* 0Λ *is* 00 *being entangled by* 11.

R abc is Euclidian flat space and R gab is anti-gravity and –R g ab is gravity and h=Planck's constant/2π *with* ρ *ab being the energy density of matter.*

2. ENTANGLEMENT AND EQUATION OF EVERYTHING

FINAL EQUATION ;$\Gamma\ abcd\ \ \phi(|n|) > \ = \ \Lambda\ \ + R\ abc + \frac{1}{2} R\ g\ ab + \frac{1}{2} e^{i\pi} R\ g\ ab \div$
$i\hbar\ \rho\ \ ab \ldots \ldots$

The Christoffell Symbol represents the derivative of the tensor of the fourth degree R abcd presenting Riemannian or Lorenzian curved space-time acting upon or being acted upon by n eigen-states of energy going from the cosmologic constant to Rayo's number or the Grand Unification energy of 10^19GEV or 10^77 joules being entangled in a non-chaotic arrangement of Spooky Action at a Distance and the interaction of strings according to type II and IIA string theory as well as the Heterotic 8x8 string theory and the SO32 string theory forming M theory formulating the Superbrane consisting of 524,288 permutations of energy levels as eigenstate. The expression R abcd(n) where n is the number of eigenstates of energy going from the Cosmologic Constant to the Grand Unification Energy acting upon or being acted upon by curved space-time being curved by the mass equivalent of the separate disparate energy states is represented by the Christoffel symbol as the derivative of the tensor of the fourth degree of space-time .The term psi(|n|)> represents the entanglement of these myriad disparate energy states acting upon each level of space-time forming the derivative of the tensor of the fourth degree of space-time. In actually Entanglement exists to some degree in every energy state except the ground state as represented by The Cosmologic Constant so when the derivative of the tensor of space-time is acted upon by the cosmologic constant the constant vanishes or becomes zero as the derivative of any constant is zero. Of course
e^πi *is* –
1 *according to Euler's Identitiy making the tensor expression for antigravity* + $\frac{1}{2R} g$ *ab(flattening space* – *time)and gravity curving space* –
time inward as the Black Holes.

Of course the ascending and descending spheres or geodesics of space-time going from a point of infinite curvature to flat space-time of infinite diameter are represented by $2\pi R$ where 2^n+1
π *is in the numerator and* $2^{n\pi}$*is in the denominator making the spiral operator acting upon an*

Angular momementum for the geodesics increasing flatness and decreasing to a point of infinite curvature.The relates to the Stress Energy Tensor time the gravitational coupling constant or
8
$\frac{\pi G}{c4}$ *T ab in the numerator and* $\frac{8\pi G}{c4}$*T ba in the denominaator. As the ground state is the cosmolo*

Cosmologic constant
or
Λ *from the First Event Time Oscillation Paradox Theory the stress energy tensors cancel leav*

Leaving R as the cosmologic constant and the expression
$2\pi\ \Lambda$ *where R is the cosmologic constant and* $2^n +$
$\frac{1}{2^n leaves}$ *2 pi times the ground state energy level whereby 2 pi is the circumference of the first*

Geodesic of space-time where the radius is the ground state energy level or the cosmologic constant.

$\Gamma\ abcd\ \phi(n) >$
$= 2\pi\ \Lambda\ +\ R\ abc + \frac{1}{2} R\ g\ ab + \frac{1}{2}\ e^{i\pi}\ R\ g\ ab \div\ i\hbar\ \rho\ ab$ *where* $\hbar$
$= h\frac{}{2\pi}$ *making the expression compactify to a circle with space*
− time as the circumference and the Riemann Forces of Nature as the positive and negative r

Radiating grom the center of the compactified circle. This is the expression space-time=space/mass.

ENTANGLEMENT;HOW FAR DOES IT GO?

Lately ,new scientific evidence(Science News;Emily Conover vol.191 #8 4/29/2017) indicates that at least millions of bits of information on a sub-molecular level become entangled with each other. This has far reaching implementations as entangled particles with data can mix or exchange information of any type anywhere.

Can this be associated with paranormal phenomena?" Mystics" who via séances can purportedly contact the dead may in a few select cases have their frontal and temporal lobes of their brain entangled with data from deceased relatives or other individuals. In my previous book "Mega-physics III; Nothing Doesn't Exist "it was explained by this author that "Spooky Action at a Distance" can show information exchange with quantum entanglement between an evaporating black hole and adjacent black holes or even black holes in other galaxies in milliseconds thus solving "The Hawking Paradox". Can ghost sitiings be related to entanglement? If memories or images from the past are stored in the hippocampus of the brain of an individual can electromagnetic impressions which are two dimensional become encoded in the memories of certain perceptive "receivers" which would subsequently be entangled with bits of information containing the spirit and manifest as a ghost siteing? Are senders and receivers all encompassed in each and every brain along the surface as indicated by the University of Arizona(microtubules which may acts as receivers and senders of K-suryon waves).

ANOTHER IDEA REGARDING THE POSSIBILITY OF TIME TRAVEL BACKWARDS OR REVERSING TIMES' ARROW

The speed of light boundary or c=3x10^8meters/sec which approximates the speed of light in a near vacuum has space-time constrict towards zero as an asymptote(space is constricted towards zero by a near infinitely long time dimension whereby time approaches stopping) acts as an interface whereby tachyons travel above "c"and fermions below "c".Above "c" these tachyons travel with a reversal of times arrow(backwards)while the hadrons which are fermions have times' arrow forward. In essence if the speed of light boundary can be breached to the upside after the constriction phase of space by time to the actual end value of the asymptote then the increase in space caused by the constriction of dilated time will have the constriction move with a reversal of times' arrow.Space-time and tachyons are the only known quantities which travel above "c" which was called as a misnomer the speed of gravity.

In essence if a large enough centrifuge effect can be achieved similar to that caused by the Time Oscillation Paradox the angular momentum may exceed the GUT or 10^19 giga electron volts. In this case the mass perturbs space in such a way that the angular momentum will appear to the intelligent observer to curve space in the constriction and dilation pattern caused by the effects of the dimension of time which is the relative speeding up or slowing down of the sequencing of events. Using the Pythagorian Calculations of angular velocities and angular momenta will produce the recorded results which match with the math showing the relative constriction and dilation of space and time. A centrifuge with a spin of greater than 2.99x10^8meters/second counterclockwise will breach times arrow causing it to receive infinite dilation after which times' arrow will reverse to times' arrow below "c". Unfortunately such G forces will splatter the time travel smoothly over the periphery of the time chamber and the heat from friction will probably incinerate what was left. With our level of technology while it may be possible to construct such a chamber surviving the trip would require safeguards which most probably are beyond our technology as it is in 2018.

How The Left Shift is Affected by the 2.74 degree kelvin Discrepancy caused by the Background Microwave Radiation from "The Big Bang"
By Dr. Mitchell Albert Wick

The Boltzmann Equation measured the entropy(S)of a state and is a fluid transport equation for matter and energy. Space- time is a perfect fluid and energy or photons act as a Perfect Gas. The Boltzmann Equation is S=k ln W where by k=Boltzman Constant or 1.38062x10^-23 joules/degree kelvin and W=number of microstates of the system. This mentions the number of ways molecules of the system in Thermodynamics can be arranged.

To find the wavelength and frequency of electromagnetic radiation at 2.74 degrees kelvin we must multiply 2.74(1.38062x10^-23)in the MKS system to factor label out degrees kelvin from the equation.
E=h
υ *accordingly the energy of a photon equals Plank'sConstant or* $6.63x10^{-34}$*joule* − sec *times the frequency of the electromagnetic radiation. Thermometers appear to be* 2.74 *deg* colder than they really are2.74(1.38062x10^-23 joules of energy in the MKS system reveal the energy of the system gained or lost by the 2.74 degree kelvin differential in measurement.E=hυ *so* $2.74(1.38062x10^{-23} = 6.63x10^{-34}(frequency\ change)$

hich is 5.706x10^11 hertz or cycles/sec in the MKS system and this is the reciprocal of wavelength
$.\lambda = 1.75438x10^{-2}$*nanomters while frequency is* $5.70572913x10^{11}$*hertz.*

Microwave frequencies are from 300 Megahertz to 300 Gigahertz so 5.70x10^11 hz is 5.70x10^5 Megahertz or 5.70x10^2 Gigahertz for the background microwave radiation from The Big Bang.Radiowaves are a longer wavelength than micrwaves going from 0.04 inches or 1 mm to 100 kilometers so the corrected frequency of 5.70x10^2 Gigahertz is the upper limit of the frequency of corrected BMR as a boundary to radiowaves which falls also in the area of UHF or ultrahigh frequency which is why "snow "appears as the BMR on a t.v.screen. As a result the distances of galaxies and stars are farther away from us than expected and the expansion of this universe is faster than projected by the uncorrected BMR.

CHAPTER 12.2 MATH continued

The peak wave length of Hawking Radiation is almost 16 times the Schwarz child Radius of a Black hole. The equation is$\lambda\, max = \left(\frac{8\pi^2}{4.9651}\right)\;\; r(s) = 15.902\, r(s) where\ r(s) is\ the\ Schwarzchild\ Radius$ of a black hole event horizon and the wavelength of Hawking Radiation relates to flat Minkowski Space-time as ds^2=dx^2+dy^2=dz^2-c^2dt^2 from the line element. With regard to Spooky Action at a Distance Hawking Radiation is electromagnetic radiation perturbing space-time in a spiral configuration so the spiral operator 2n+1^$\pi\omega\, i \rightarrow j(for\ expanding\ space - time\ from\ a\ point\ to\ flat\ space - time\ for\ Hawking\ Radiation\ to\ perturb\ space - time\ toward\ the\ adjacient\ black\ hole\ event\ horizons\ where\ space - time\ constricts\ by\ the\ spiral\ operator\ to \frac{1}{2^{n\pi\omega}\, j} \rightarrow iand\ the\ angular\ momentum\ increases\ geometrically\ as\ the\ event\ horizon\ of\ \ all\ other$

black holes are approached. The energy of Hawking Radiation is $\rho(H)\;\; in\ the\ Equation\ of\ Everything\ is\ \ i\hbar\rho(H) so\ that\ 2\pi(R\ abc + or - \frac{1}{2} R\ g \frac{ab}{i} h\rho(H) and\ \ \Sigma\rho(\mathrm{H}) = \lambda\, \mathrm{max}\ or\ the\ peak\ wavelength\ which\ is\ the\ reciprocal\ of\ the\ frequency\ \upsilon\ which\ relates\ to\ the\ en$

energy of Hawking Radiation. As photon pairing is electromagnetic radiation and is subject to Spooky Action at a Distance photons also follow the paths of space-time curvature in the same way as other types of electromagnetic radiation as all are comprised of photons including Hawking Radiation.

Utilizing the formula for space-time as -1/2e-I n cot(theta) where theta is the trajectory of Hawking Radiation the trajectory is 90 degrees or $\frac{\pi}{2} radians\ which\ is\ \cos \frac{\frac{\pi}{2}}{\frac{sin\pi}{2}}\ \ which\ is\ \frac{0}{1} or\ 0\,. This\ is\ consistant\ with\ the\ constriction\ of\ space - time\ towards\ zero\ at\ the\ event\ horizon\ of\ any\ active\ black\ hole. Thus = \frac{1}{2} e^{-in(0)w} heren\ is\ the\ two\ dimensional\ state\ relating\ to\ the\ flat\ matter\ of\ Hawking\ Radiation$

e^0=1 and 1/2e^0 =1/2 while -1/2e^0=-1/2 so the midpoint is the inception of a self contained LOOP OF HAWKING RADIATION IN CONSTRICTED TO DILATED SPACE-TIME FROM THE EVENT HORIZON OF ONE BLACK HOLE TO THE EVENT HORIZON OF ANOTHER BLACK HOLE.THIS IS TOTALLY CONSISTANT WITH TOTAL INFORMATION EXCHANGE OF HAWKING RADIATION BETWEEN EACH AND EVERY BLACK HOLE WHERE TIME ISN'T IMPORTANT AS HAWKING RADIATION GOES FROM CONSTRICTED SPACE-TIME OF ONE BLACK HOLE WITH DILATED TIME TO CONSTRICTED SPACE-TIME WITH DILATED TIME OF ANOTHER BLACKHOLE THEREFORE THE INFORMATION DISBURSED INTO SPACE-TIME AS EITHER A SUSPENSION(where the 2 dimensional lattice formula can be applied)or if the information is dissolved in space-time and may travel in a formulation similar two the Robinson Congruence (which is a superfast pathway for photons which comprise electromagnetic radiation).These phenomena also explain photon pairing as a form of Spooky Action at a Distance .The phenomena involving spooky action with electrons was described in this author's second book "The Equation of Everything". 12.3 AN M THEORY APPROACH TO HAWKING RADIATION Assuming that Hawking Radiation is comprised of electromagnetic radiation or photons which are flat or two dimensional matter this would be associated with the 3-Brane associated with two dimensional space ,the 4- Brane associated with three dimensions of space where the third dimension of photons would be string sized or 10^-33 cm. The gathering or distribution of charge in the 2-Branes are generally associated with electrons which also has electromagnetic radiation as magnetism is formed from electromagnetic radiation as electrons are transferred as well as positrons(anti-electron)and the fact that photons have a miniscule mass would cause them to smear along multiple membranes or Branes as the velocity of electromagnetic radiation would only vary at just above absolute zero when photons vibrate in a lattice of Boso-Einsteinian condensate. Space itself would be comprised of leptons and gluons and photons would pass through the fabric of leptons as long as the gluons remained intact holding the leptons together and would therefore form the 4-Brane,5-Brane,....N-Brane all the way to the super-small Super-brane There were 252 separate states of matter in the average active black hole according to Stephen Hawking and black hole entropy

S=2

$$\pi\sqrt{NQ1Q5} \quad where\ N = number\ of\ states\ Q1\ and\ Q5\ are\ the\ differential\ charges$$

.Spooky Action between black holes would have space altered such that the 3 Brane would compress toward the superbrane constricting the volume of photons which would increase the volume geometrically BY THE MUZZLE EFFECT AS WITH A WATER BLAST IN A MUZZLE.CONSIDER ELECTROMAGNETIC RADIATION AS A PERFECT GAS WITH THE PERFECT FLUID OF SPACE-TIME;CONSTRICTING THE PERFECT GAS ENOUGH WILL INCREASE THE VELOCITY TO EXCESS OF 3x10^8 meters.sec as the velocity of space-time would also increase geometrically with constriction nd the photon stream of Hawking Radiation would constrict with the constricting space-time.Therefore photons would constrict to sizes which are virtually immeasurably small and immeasurably fast in the massive Super-brane explaining Spooky Action at a Distance for Hawking Radiation using M Theory.

FOOTNOTES;

1.Kaku ,Michio. Introduction to Superstrings and M Theory p.61-62 and section 1.8 Harmonic Oscillator

2.Peebles.Principles of Physical Cosmology p.500-503

3.Wikipedia.Hawking Radiation.

CHAPTER 13:

WHAT IS MORE LIKELY THAT THE BIG BANG WAS PRECEDED BY A BIG CRUNCH AND WAS NOT THE SECOND EVENT BUT HAPPENED AEONS LATER OR THAT THE BIG BANG WAS THE SECOND EVENT FOLLOWING THE FORMATION OF STRINGS,STRING DIMENSIONS AND MATTER?

Utilizing Occam's Razor the first explanation appears on the surface to be more plausible as it is logical and simple ,however ,is it? The second explanation would place the site of the Big Bang at the center of the vortex formed by the Time Oscillation Paradox acting on the infinite parallel planes of space. In order for the second explanation to be more correct would have to illustrate a recoil effect or spring effect in the opposite direction from the Big Bang which initially went outward in all directions from the center of the spiral vortex of space-time which had been previously formed. This spring effect would have had a deceleration early on after The Big Bang after which time the acceleration outward in all directions would have been a pull from space-time acting on space-time from our universe as a current along with the push of dark energy which is a relatively weak force of anti-gravity. In other words there was a push outward from Dark Energy from the mutual repulsion of anti-particles in the quantum bubble from the center of the vortex but this push was almost IMMEDIATELY overtaken by the pull of space-time back toward the center of the vortex causing the accelerated expansion toward the center of the vortex .Note immediately could be anything from Planck Time or 10^-43sec to several hundred thousand years. The answer is in the data .If the WOMP or any studies of gravity waves or measurements of space-time curvature would show a change or reversal in the acceleration early on followed by a substantial increase in the acceleration .As a consequence ,with the knowledge our technology has the conclusion must be soft or tenuous at best albeit possible. Theo other alternative which is simpler and easier to explain is a Big Crunch of another universe with space-time constricting toward zero and time dilating toward infinity (infinitely slow relative to the observer) which would mean the progressive dilation of time would slow either slow it toward zero from positive time moving progressively slower until time 0 is approached as the quantum bubble is formed in constricted space-time or time would move backwards from positive toward zero in a negative direction as in the case of a massive IMPLOSION. In this case The Big Bang was certainly not the second event but occurred much much later. Of course the collision of membranes could also have caused it but of course this too would mean it isn't the second event but would have happened much much later. So what is correct? Are all things equal? If science can answer that question we will be considerably closer to the OMEGA POINT. To learn all things that are learnable would answer the question "When are all things equal? ".Until then the first statement toward wisdom is "I don't know".

13.2 ANTIPARTICLE ANTIPARTICLE REPULSION AS CAUSE OF THE BIG BANG

IN A SYMMETRICAL 360 DEGREE ORB BLAST AT THE BIG BANG R^ijkl=antigravity effect on particles and Rji^kl is the antigravity effect on antiparticles.Λij $= 1/Rij$^2 =$1/8\pi G$ $where\ R\ ji =$ $antigravity\ from\ antiparticles \frac{1}{Rij^2 is} spacetime\ from\ the\ Cosmologic\ Constant. The\ vector\ prod$

Rij^2⊗$R\ ji\ reveals\ R^{2ijRji}/|Rij^2||Rji =$
$cos\theta\ which\ revelas\ that\ e\ ji\frac{R^{2ijR}ji}{||R^{2ij}||}||Rji||\ where\ 0 < \theta <$
$\pi\ radians. In\ a\ symmetrical\ orb\ blast\ the\ re\ are\ an\ infinite\ number\ of\ slices\ with\ a\ total\ trajec$

tory of
$\pi\ radians. The\ cosine\ of\ \pi\ radians\ is - 1. so\ R\ ji =$
$\frac{1}{8\pi G\otimes g}ij\ The\ vector\ product\ \ of\ Rij(R\ ji) = e(ij,ji)cos\theta. Rij^{2(R\ ji)} =$
$1. For\ \pi\ radians\ R\ ji = -e\ ji. As\ \ g\ ij\ is\ the\ metric\ of\ a\ particle\ and - eji -$
$is\ sspacetime\ \ curvature\ or\ gravity\ of\ R\ ji. g\ ji\ is\ the\ metric\ of\ the\ antiparticle. R\ ijkl\ is\ spacet$

Time for antimatter in four dimensions and kl is positive space-time. With R jikl=covariant tensor and Rji^kl being the contra-variant tensor R ji=-e ji and R jikl-Rji^kl=-eji=g ji/8(pi)G where R jikl-R ji^kl is the space-time curvature from anti-matter,R jikl-R ji^kl=-
$8\pi G\ eji = g\ ji\ where\ g\ ji\ is\ the\ Riemann\ metric\ for\ an\ antiparticle\ and -$
$8\pi G\ is\ the\ ANTIGRAVITYFOR\ ANTIPARTICLES\ IN\ ANTIMATTER\ ILLUSTRATING\ A\ MUTUALLY\ R$

REPULSIVE FORCE.R ijkl is space-time curvature for gravity and R jikl is space=time curvature or reciprocal curvature for antigravity and of course g ij is the metric for particles with positive gravity and g ji is the metric for the anti-particle with anti-gravity. Finally R ^ijkl is the antigravity effect on particles and R ji^kl is the antigravity effect on antiparticles which forms the -
8
$\pi G\ which\ is\ the\ multiplier\ of\ \ e\ ji\ to\ equal\ the\ metric\ g\ \ ji\ which\ is\ the\ metric\ of\ the\ antiparticle$

Clearly the negative sign illustrates anti-gravity for anti-particles.

13.3:EVERY ACTION MUST HAVE AN EQUAL BUT OPPOSITE REACTION

According to Sir Isaac Newton every action must have an equal but opposite reaction. The Big Bang ex nihilo breaks that law of motion; however with the spring effect of the quantum bubble from emerging strings and string dimensions from the center of the vortex of spiral space-time IS THE ACTION after the time oscillation paradox spuming out in all directions to the point of deceleration from the center of the vortex. This then reverses toward the center of the vortex again with the accelerated expansion of our universe. Again this is due to the enormous pull of space-time from the vortex on space-time and push of dark energy within our universe and is the REACTION.. The equal but opposite reaction to the expansion of our galaxies from the quantum bubble and Big Bang was the push of anti-particle anti-particle repulsion within the quantum bubble from the center of the vortex to the POINT OF REVERSAL OF THIS UNIVERSE BACK TOWARDS THE CENTER OF THE VORTEX OF SPACE-TIME. Of course a Big Crunch of another universe with the same space-time current as our universe would also obey Newton's Second Law.

Therefore the event of "creation" in all actuality occurred when the rogue tachyon (s) dropped to the speed of light initiating the time oscillation paradox on the infinite parallel planes or the D-0-Branes. Again it is unclear as to whether only one universe or the multiverse formed from the advent of string dimensions and strings as well as the spin 2 vector boson from the rogue tachyon(s)initiating gravity and converting almost 100% potential energy of the components of space into 10^77 joules of kinetic energy after the paradox acted upon these infinite parallel planes spinning them into a centrifuge effect forming the vortex of space-time with matter gravitating towards the center as an origination point. THIS IS CLEARLY NOT OUT OF NOTHING AS NOTHING DOESN'T EXIST.

13.4 WAS THE ROGUE TACHYON COMPONET OF THE HIGGS FIELD (PRIOR TO THE FORMATION OF THE MASSIVE HIGGS BOSON) A PURPOSEFUL ACTION?

To mathematically determine if the rogue tachyon (s) was impelled by a conscious force(determinism vs. free will) is bringing together science and philosophy merging into religion and while K-suryon waves were determined to mathematically exist; it still hasn't been conclusively proven that consciousness is energy. Until that question is answered it will be unclear if creation was purposeful or random(which according to Quantum Mechanics would have a 100% probability with infinitely or nearly infinitely dilated time) for that event to occur. Chaos Theory would indicate randomness which follows the laws of entropy which would indicate that the most ordered state would be prior to the first event. Again if the vortex of space-time relates to the vortex of a black hole then the entropy would approach zero in the center of the vortex but not be zero. Stephen Hawking stated that black hole entropy approaches 0.29 and it is possible that the entropy of the quantum bubble would also approach that value in the center of the vortex of space-time with 252 disparate states of matter and space-time constricting towards zero without reaching it at the center(BEING THE LOCATION OF THE BIG BANG).

13.5 THE DISTRIBUTION OF SPACE-TIME AFTER THE FIRST EVENT

Space-time acts as a perfect fluid and the spiral configuration of the space-time continuum is like a giant whirlpool whose outer edges stretch out toward infinity in all directions from the center of the vortex. As the distance from the center of the vortex increases in all directions over 360 degrees the amount of curvature reduces until at the extremes (which are never met) space-time approaches absolute flatness which is like a stagnant pond. The quantum ground state as described by the spiral fractal formula is the ZERO DIMENSIONAL STATE and 100% potential energy-
$\epsilon(negligible\ kinetic\ energy) is \sim zero\ energy\ but\ is\ actually\ \varepsilon\ kineitic\ energy\ and$ 100% $-$ $\epsilon\ potential\ energy\ from\ the\ components\ of\ space.$ $\int_0^{\pi} \frac{du}{u} = \ln u \rightarrow 0$.VACUUM ENERGY IS THE SUM TOTAL OF THE POTENTIAL ENERGY OF SPACE WHICH IS COMPRISED OF LEPTONS AND BEING HELD TOGETHER BY MASSLESS GLUONS.IT CAN BE CONSIDERED AS 10^77 joules incurred by the Time Oscillation Paradox of the first event. Of course as mentioned in this author's previous book "What is the Dimension of Time?" time is the sequencing of events and without time all events would occur simultaneously in the same space which absolutely disallows space-less-ness as space-less-ness cohabiting space containing everything with sequencing

would implode everything into space-less-ness which would require time as an implosion is an event as would be the formation of space from space-less-ness therefore they would have been nothing without time which can only be nothing and could have only been nothing WHICH IS NON-EXISTENCE therefore once again nothing doesn't exist and time does.

The classical definition of potential energy is P.E.=m g h where m=mass g=gravity and h=height .In the infinite planes hypothesis the height of the infinite planes is infinity. The mass of leptons is approaching mass -less-ness but still has a miniscule negligible mass and therefore negligible gravity which bleeds through from tachyons. Gravity is the curvature of space-time caused by mass ,but the curvature of space is from time constricting or dilated space. If time is infinitely or nearly infinitely dilated the constrictions of space by time approaches ZERO so the pre-first event space is flat and totally parallel .As a consequence the negligible mass causing gravity cannot curve space with time completely dilated as it is time that is actually constricting space. As a result

mass=ϵ $gravity = \epsilon$ $and\ space - time\ curvature =$

$0\ as\ time\ is\ dilated. So\ the\ potential\ energy\ is\ (\epsilon)(\epsilon)(\infty) =$

$\infty\ and\ there\ is\ therefore\ \infty(infinite) potential\ energy\ which\ released\ 10^{77 joules} of\ kinetic\ energ$

from the first event. Of course the Newtonian Definiton of Potential Energy as weight times height where weight is mass times gravity is only an approximation as it may not consider Relativistic effects on potential energy.

Type equation here.CHAPTER SIXTEEN CONTINUED

THE EFFECT OF DARK MATTER AND DARK ENERGY IS LIKE LIGHTING A MATCH WHERE THE PHOSPHORS ARE THE ANTIMATTER AND MATTER ,STRIKING THE MATCH IS THE BIG BANG,AND DARK MATTER IS THE CHARRED RESIDUE OF THE MATCH WHILE THE FLAME IS DARK ENERGY(FROM ANTIMATTER)and HEAT(FROM MATTER PLUS THE EXPLOSION FROM MATTER ANTIMATTER ANNIHILATION)WITH ANTIGRAVITY FROM ANTIMATTER ANTIMATTER REPULSION.AS DARK MATTER IS VERY LOW ENERGY IT's COMPONENT PARTICLES MUST BE VERY LOW ENERGY AND THESE PARTICLES MAY BE THE NEXT ENERGY LEVEL ABOVE THE VACUUM STATE IN THE MASS GAP OF THE YANG MILLS THEORY.THE$\Omega -$
$PARTICLE\ MAY\ BE\ AN\ ELEMENTARY\ PARTICLE\ IN\ DARK\ MATTER\ BUT\ IT\ MUST\ HAVE\ CHARA$
CHARACTERISTICS IN COMMON WITH QUARKS IN THE SU(3)SUBGROUP.QUARKS AND GLUONS KEEP NUCLEII TOGETHER IN ATOMS AS DARK MATTER IS PURPORTED TO ACT AS COSMIC GLUE LIKE GLUONS. The mass gap in Yang Mills must be as pervasive as the missing mass in the universe for this to hold. The mass gap is between the vacuum state and the next lowest energy level with the smallest mass possible for a particle. If dark matter has so many homogeneous particles that they cannot be directly detected ,these particles may be so small that they meld or mesh into space-time or in the case of Yang Mills Euclidian space. Based on this it's possible that dark matter would most likely account for the mass gap in Yang Mills Theory although it would occur in multiple gauge symmetry groups but least in U(1)as they have little in common with electromagnetic radiation .With regard to electroweak forces of nuclear decay and gravity the dark matter particles may form an **integral** part of the U(1)x S(2) and S(2)gauge symmetry groups .Being non-abelian though they must share reflective symmetry patterns with other members of these groups.The temperature of space is 2.74 degrees kelvin .Dark Energy is so pervasive in space that unlike other energy it must be COLD.**Scientifc measurement can not verify high energy with Dark Energy. Dark Matter is extremely low energy and possibly extremely compact small mass particles which may stick or adhere to other particles.In terms of gauge symmetry the grouping would be U(1)xSU(2)xSU(3)x...SU(n) as it is likely associated with electroweak forces,gravity and The Strong Force.These microparticles may be the**$\Omega -$
$particle\ in\ SU(3) and\ may\ relate\ to\ all\ matter\ except\ for\ photons. Also\ photons\ and\ electromag$

electromagnetic radiation can be of very high or lower energy levels and the energy level of a microparticle of dark matter may be below that of a cold photon.The mass of this microparticle would be the total mass of dark matter or 3.3x10^24 kg/number

of particles per volume of the universe where the volume of the universe is derived from

$\rho c(critical\ density = mass\ of\ the \frac{universe}{volume} of\ the\ universe. So\ \rho c = mass/volume$

and from Poisson's Equation

$\mathbf{4}\pi\rho = \nabla\left(\frac{mass}{volume}\right) where\ mass\ and\ volume\ are\ the\ dual\ vector\ field. Mass\ can =$ $the\ Ricci\ Tensor\ R\ ab\ and\ volume\ is\ \mathbb{R}4\ space. Therefore\ 4\pi\rho = \frac{Rab}{\mathbb{R}} = R\frac{ab}{Rabc} -$ $R\ g\frac{ab}{R}\ ab\ or\ the\ vector\ product\ of\ R\frac{ab}{Rabc} - \frac{1}{2R}Rg\ ab + \frac{1}{2}R\ g\ ab$ **where R g ab is antigravity and –R g ab is gravity suggesting reciprocal curvature and curvature of spacetime**$\mathbb{R}.R\ g\ ab >$ $-R\ g\ ab\ as\ antigravity\ causes\ H0\ and\ the\Omega\ of\ the\ universe. Therefore\ \ 4\pi\rho =$ $R\ ab(as\ a\ vector\ product\ of\ R\ ab.R\ ab) - 0 +$ $\frac{1}{2}R\ g\ ab\ as\ antigravity\ exceeds\ gravity\ due\ to\ H(Hubble\ expansion\ at\ time\frac{t}{H0}Hubble\ expansion$ **at time 0. So it follows that** $4\pi\rho = R\ ab.R\ ab\frac{1}{2}R\ g\frac{ab}{R}a\ b\ c\ or\ R\ ab.R\ ab +$ $\frac{1}{2}R\ g\ ab \div R\ abc\ \ or\ \rho = R\ ab.R\ ab + \frac{1}{2}R\ g\frac{ab}{4\pi}R\ abc\ or\ \rho = R\ ab' + \frac{1}{2}R\ g\ \ ab \div$ $4\pi R\ a\ b\ c\ \ wheree\ R\ ab' = R\ ab\ .R\ ab.$

R ab=3.3x10^24 kg for Dark Matter Rab'=10 ^54 kg for the universe

$\rho c = critical\ density\ of\ the\ universe\ and\ \rho = density\ of\ dark\ matter. \frac{\rho}{\rho c} =$ $R\ ab \div R\ ab.R\ ab = 1 \div R\ ab\ \ which\ is \frac{1}{3.3x10^{24}kg}or\ .33x10^{-24} =$ $\frac{\rho}{\rho c}which\ approaches\ 0\ density\ for\ dark\ matter\ yet\ it\ has\ extensive\ mass\ which\ is\ in$ **in line with scientific observations.The particles would be so small that they would need to be compactified to 0 (curled up)with 0 density,positive mass and positive gravity.As particles or microparticles of dark matter→** $\infty\ they\ may\ be\ below\ Planck\ length\ 10 - 33\ cm\ in\ size\ and\ very\ low\ in\ energy$

causing the mass gap$\Delta > 0\ in\ Yang\ Mills\ Theory.$

The Hubble expansion is such that $H^2 = \frac{8\pi}{3m^{2\left(1+\frac{\rho}{2\vartheta}\right)}} + \Lambda 4 + \frac{E}{a^4 or}\ or\ 8\pi\left(1 +\right.$ $\left.\varrho \div \frac{}{2}\sigma\right) + \Lambda 4 + \frac{E}{a^4 w}herem =$ $masss, \rho\ is\ the\ energy\ density\ of\ matter, e\ is\ energy, a\ is\ cross\ sectional\ areaHubble\ expansion$ $\left(a.\frac{}{a}\right) or\ a' \div a = \frac{8}{3\pi G\rho e}heremass\ \ density\ is\ \rho(x,t) == \rho b(t)\{1 + \delta(x,t)\}so\frac{a'}{a} =$

$\frac{8}{3\pi G\rho B}. or\ 8 \div$
$3(\pi G)(\rho B) where \frac{8}{3} is\ multiplied\ by\ \pi G\rho B. and\ the\ time\ evolution\ of\ the\ expansion\ parameter$

I(as a(t) where a is the expansion parameter in the expression a./a or a'/a with a being the area being expanded.$\frac{a(dot)}{a})^\wedge 2 =$
$8/3\ \pi G\rho B + 1/a^\wedge 2R^\wedge 2 + \Lambda/3. a(t) = a0[1 - H0(t0 - t) - q\frac{0Ho^{2(t\ 0-t)^2}}{2} + \cdots or\ [1 -$
$H\ 0(t\ 0 - t) - q\ 0H0^{2(t\ 0-t)^2} \div 2where\ t\ 0 -$
$t\ is\ the\ lookback\ time\ and\ the\ universe\ based\ on\ redshifts\ is\ H0t0 =$
$H0\int_0^{a0}\frac{da}{a}. so\ the\ spiral\ fractal\ formula\ \int\frac{du}{u} applies\ for\ the\ expansion\ in\ space -$
$time\ and\ is\ based\ on\ red\ shifts. \int_1^\infty dy/y[\Omega y^3 + \Omega(R)y2 + \Omega(\Lambda]1/2.$

$$\Omega(R) = \frac{1}{(a\ 0\ H\ \ 0\ R)^2 and\Omega}\Lambda = or\Omega\left(\bigwedge\ \right) = \Lambda/(3H\ 0)^\wedge 2.$$

$\rho c = \frac{3H^2}{8\pi G} and\ \Omega = \frac{\rho}{\rho c} = 8\pi G\rho/3H^\wedge 2$ **Peebles's 23**

As a perfect fluid
p$\Lambda = \frac{\Lambda}{8\pi G} and\ p\Lambda = -p\Lambda\ and\ 4\pi G = \Lambda =$
$\frac{1}{R^2 w} herethe\ energy\ density\ of\ matter\ causes\ pressure\ gradients\ or\ slopes$

So
H^2=8$\frac{\pi}{3m^{2\left(1+\frac{\rho}{\sigma}\right)}} + \Lambda\frac{4(in\ 4\ space)}{3} +$
$\frac{E}{a^4} has\ \ \sigma relate\ to\ the\ pressure\ gradient\ on\ the\ energy\ density\ of\ matter\ from\ the$

Expansion of space-time based on the Friedmann Equations.Rewritten H^2=8$\frac{\pi}{3m^{2(1+\rho\div 2\sigma)}} + \Lambda\, 4 \div 3 + E \div a^\wedge 4.$
H^2=8(pi)G/3m^2(!+p(rho)/2(sigma)+cosmologic constant in Riemann 4 space/3+energy/a^4

Einstein's Perfect Fluid Equation is R^ij=k(T^ij-1/2g^ijT)+$\Lambda\ g\ ^\wedge ij$

T^ij is stress energy$\kappa = 8\pi. k = Gaussian\ curvature\ in\ length\ ^\wedge 2.$

σ *relates to* $\frac{\rho^2}{\sigma}$ *whose effect is to increase dampening experienced by the scalar field*

as it rolls down its potential

where$\rho = \frac{1}{2\phi^2} + V(\phi)$ *and the potential is* $V(\phi)$ *so* σ *is the gradient of the field*

ρ
$= 0.5\phi^2$
$+ V(\phi)$ *is the expression for the energy density of matter with regard to the scalar field* ϕ

The initation of the Big Bang with a rotational vector in the quantum bubble

An infinitesimal roation through a small angle

d

θ *about a rotation axis with the direction cosines* $c1, c2, c3$ *is described by the orthogonal infin infinitesimal transformation* $x' = (l + \Delta)x$ so that

Δ *is skew symmetric. With a reference frame with elements* $a1, a2, a3$ *which are orthonormal*

Mutually perpandicularΔ *is represented by the skew symmetric matrix* $\Delta =$

$$\begin{matrix} 0 & -a3 & a2 \\ a3 & 0 & -a1 \\ -a2 & a1 & 0 \end{matrix}$$

for$\theta = d\theta \to 0$ *such that* $\Delta x = (a \; x \; X)d\theta$ *where* $a =$
$a1u1 + a2u2 + a3u3$ *which is a unit vector in a* $+$
rotation axis. For a continuous D3 rotation at time 0 $x'(t) = A(t)x$ *where* $x =$
constant giving $\frac{dx'(t)}{dt} = \frac{dA(t)}{dt}$ *times* $x = \omega(t)x \; x'(t) =$
$\omega(t)x \; [A(t)x]$. ω *relates to angular velocity and the vector* $\omega(t)$ *is given with reference to fixea*

and rotating vectors

$\omega(t) = \omega1(t)u1 + \omega2(t)u2 + \omega3(t)u3 = \omega1(avg)(t)u^{1'}(t) + \omega2(avg)(t)u'2(t) +$
$\omega3(avg)(t)u'(t)$ **direcected along the instantaneous axis of rotation(axis x' to x'+dx') and**

$\omega(t)|$ *or* $|\omega(t)$ *is the instantaneous absolute rate of rotation.* **It follows that aA(t)/dt=**

$\Omega(t)A(t)$ *with* Ω *as the skew symmetric operator with reference to* $u1, u2, u3$ *and* $u'(t), u2'(t)$ *anc*
u3'(t) as elements on which the skew symmetric operator with respect to time works as a 3x3 matrix for$\Omega(t)$ *and* $\Omega(avg)(t)$ *so if* $A -$

$1\ power\ is\ called \frac{A(\sim)dA}{dt} = \Omega A = A\Omega(bar) and \Omega = \frac{dA}{dt} A(bar) and\ \Omega(bar) = A \sim$ $\frac{dA}{dt}$ or $\frac{A^{-1}dA}{dt}$. *the three dimensional Rotation Group – relection group R3 ± has*

proper rotations [det(A)=1] which is a proper rotations group and is a subgroup of R3+- the three dimensional rotation group R3+.R3+- and R3+ don't follow the communitive law . Space-time has constricted space curved by mass of stringx7.5x10^11 strings from infinite curvature to near flat space-time.$\mathbb{R}4 \longleftarrow \mathbb{R}3$ *where* $\epsilon(t) \leftarrow 0. \int_{\infty}^{0} \frac{du}{u} = \ln\infty - \ln 0 = \infty$ *curvature for*

A point with infinite curvature at the point of maximum curvature And approaching 0 curvature as space-time approaches flatness.At $\mathbb{R}3$ *the three dimensional rotation group applies with dimension* **N=2j+1 and a 2j+1 by 2j+1 matrix represents the rotation of Cayley-Klein parameters. In this case j=1 for N=3 and the unitary irreducible representations are** $\mathbb{R}\frac{1}{2}, \mathbb{R}1, \mathbb{R}\frac{3}{2}, \mathbb{R}2, \mathbb{R}\frac{5}{2}, \mathbb{R}3$*withCayley – Klein parameters* $a, b, -b*, a*$ *and Euler Angles* $\alpha, \beta, \gamma. \mathbb{R}j = xj(\alpha,\beta,\gamma) = Tr[U^j mq(\alpha,\beta,\gamma) =$
$\sin(j+\frac{1}{2})\delta / \sin\delta(.5)$ *where* $j = \frac{\frac{0,1}{2,1,3}}{2}, \ldots)$ *where* δ *is angle of rotation.* $j = .5,1.1.5$ *and* $m, q = -j, -j+1, -j+2 \ldots j-2, j-1, j.$ *Tr is the trace(spur. of the* $2j+1 \otimes 2j+1$ *matrix.*

The above is spherical surface harmonics of degree j. For integral values of j the functions represent a (2j+1)dimensional representation for
space
$\mathbb{R}(j)$*with orthonormal functions. Direct products of* $\mathbb{R}3$ *rotation groups describe the* **Composite rotations of dyanamic systems such as the pre-Planck time quantum bubble.The Clebsch-Gordan Equatiom is** $\mathbb{R}^j \otimes \mathbb{R}^{j'} = \mathbb{R}\hat{}(j+j') \oplus \mathbb{R}\hat{}(j-1) \otimes \mathbb{R}\hat{}(J'-1)$=$\mathbb{R}\hat{}(j+j') \oplus \mathbb{R}\hat{}(j+j'-1) \oplus \mathbb{R}\hat{}(|j-j'|)$
APPLY THE ROTATING VECTORS WITH THE SKEW SYMMETRIC OPERATORΩ *FOR* $J = 1$ *or three dimensions of Riemann space* **with vectors**$\omega 1(t), \omega 2(t) \omega 3(t)$ *where* $t \leftarrow 0$ *using* **u1,u2,and u3 in a 3**$\otimes$ 3*matrix.***Applying the mass of a string** $2\pi T \Sigma n = 1\ to \infty \Sigma i = 1\ to\ D - 2\ \alpha - n^i \alpha n\hat{}i$ **where**

$\alpha^1 = Regge\ slope\ and\ 2\pi T \rightarrow$
$\frac{1}{\alpha(one)} gives\ the\ square\ of\ the\ mass\ of\ an\ open\ string\ in\ two\ dimensions\ and\ is\ the\ squared$
Mass operator acting on the composite rotations in three dimensions for Riemann space at time time 0 to time t.This value reflects almost infinite curvature of spacetime with 7.5x10^11strings and the mass operator of an open string reflecting the Ricci Tensor for open strings.n is the integer valued excitation level of the quantum harmonic oscillator which for closed strings m^2 has different oscillations with regard to n(2
$\pi T\ with\ a\ standing\ wave\ function\ and\ with\ two\ chiral\ components\ (mentioned\ previously)$
Left and right handed moving oscillators going from the fermionic vaccum state toward the boson state as the mass curves Riemann space and uncurves Riemann space where the curvature is increased by gravity of matter and uncurvature is increased by antigravity of antimatter.The Dark Energy pushes out the quantum bubble from its rotational state to an expansive state in 10^-43 seconds as the infinite curvature of space-time uncoils to almost flatness.The ends of the quantum bubble(polar regions)rotate in opposite directions one metric g a b for matter provides curvature of spacetime and other metric =-g ab for antimatter forming a twisted double torus with the center being the weakest point where "The Big Bang"outward pressure is pushed.R g ab=-R g ba for gravity effect with spacetime curvature metric.For antigravity the antimatter has a curvature variant of R(-g)ab=R (-g)ba=-R g ab.The infinite momentum limit(R)relates as the reciprocal of the Neven-Schwarz 5-brane 1/2NS -NS 2 potentials relating to the fermionic or vacuum state with D-0-Branes.R ab is the mass of all strings and the angular velocity$\omega 0 \rightarrow \omega t\ gives\ a\ centripedal\ force\ 2\pi T\Sigma\omega 0 \rightarrow \varpi\ t\Sigma\ 1 \rightarrow$
$D2\alpha\ \omega 0\ \alpha\ \omega\ t \rightarrow 0 = m^2 operator\ on\ \ \omega\ and\ r =$
$radius\ of\ quantum\ bubble\ is\ 10 -$
$33cm\left(7.5x10^{11 strings}\right) giving\ radius\ 7.5x10^{-22} cm\ with\ massof\ string(\varpi\ t)^2 \div$
$r\ makes\ centripedal\ force\ 7.5x10^{22} erg(\varpi\ \ t)^2 m^2. R = m\varpi \rightarrow$
$0\ as\ massmof\ a\ string\ \ \rightarrow 0\ so \frac{1}{R} \rightarrow \frac{1}{0} \rightarrow \infty or\ the\ infinte\ momentum\ limit. D -$
$0 - branes\ go\ to\ D - 5\ branes\ as\ uncoiling\ of\ spacetime\ by\ antimatter\ occurs.$

CHAPTER EIGHTEEN

BLACK HOLES AND RELATIONSHIP WITH DIFFERENT STATES OF MATTER AND BLACK HOLE ENTROPY

Matter and energy are entangled under some specific circumstances .Are strings flat matter which is 2 or possibly 1 dimensional or are they 0 dimensional as energy only with motions tension and vibrations as only frequencies.Is there a Law of Conservation of Matter as matter and energy are inter-changible and entangled under the extreme pressures and temperatures in a black hole? According the Schwarzchild Space-time there is a constriction of space-time with time dilating to almost infinity or in essence stopping at the Event Horizon of a Black Hole. Black holes emanate from collapsing matter in galaxies, neutron stars and possibly universes. As a consequence there should be a black hole at the 0,0 point which is the point of the "Big Bang" although in an isotropic universe scientists may not be able to locate it for a considerable period of time The initial rotational vectors in the expanding universe where rotation is progressively decreasing as space-time continues to push outward should be slightly measurable as the proximity to this 0,0 black hole is approached although iso-tropism generally precludes a center of gravity. Still it makes sense that the 0,0 point would be a center of rotation for the rotating and expanding universe(this universe in the multiverse)It is now postulated that all galaxies have a central black hole including the Milky Way and these galaxies rotate along the central axis of these black hole albeit at an extremely slow rate.If negative mass exists in a black hole it was mathematically determined in chapter of Schwarzchild Space-time that superimposition of region 3 on region 2 and regions 1 and 4 can cause the information to garner at the Event Horizon like a phonograph record or video tape .This relates to the idea of The Holographic Universe and translates everything into two dimensions .Regions 1 and 3 are left and regions 2 and 4 are cancelled. Region 3 has negative mass and superimposes on region Region 3 is reflected back on region2 with region 3 having a negative mass and region 2 having a positive mass and there canceling as per the math in chapter 5,leaving regions 1 and 4.This may be a solution to the Hawking Paradox which says that as a black hole evaporates information in the black hole is lost.

Regarding the 252 different states of matter within a black hole(Hawking) these would have to include all or most state under extreme pressure with significantly constricted space and extremely high density .If electromagnetic radiation including photons from light are absorbed into a black hole(making it invisible)would that radiation show matter or matter like characteristics such as the states of liquid radiation suspended in a condensate as in the experiment by Dr. Len Hau mentioned in the previous chapter .If radiation could occupy the different states of matter and if matter and radiation can be or are entangled then radiation can occupy a liquid and possibly a Boso -Einsteinian Condensate state at near 0 degrees kelvin. If this is true perhaps radiation can in the future be considered a "perfect gas" but again for that one would have to demonstrate mass in a photon although of course electrons and photons have mass as do anti-protons and positrons although the mass in the latter two may be considered "strange mass" which might be an oscillating hybrid between negative and positive mass although recorded as positive mass which may or may not have anti-gravitational effects rather than the effect of pure gravity.

In terms of The Equation of Everything" as space-time curves in manner reciprocal to ordinary space or curves inward when in the expanding universe space-time curves

outward$\mathbb{R}n = R\ abc - \frac{1}{2}\ R\ g\ ab \div R\ ab\ or \mathbb{R}n = \prod n = 1\ to \infty \frac{1}{2^n \pi} g\ ab\ R\ abc -$

$\frac{1}{2} R\ g\ ab \div$

$\rho\ ab\ where\ the\ infinite\ product\ of \frac{1}{2^n \pi}\ times\ the\ metric\ g\ ab\ times\ Mintkowski\ Space -$

$time\ divided\ by\ the\ energy\ density\ of\ matter\ for\ the\ metric\ g\ ab\ reveals\ for\ a\ large\ mass$

In constricting space reaches a point where -1/2R g a b=R a b where the space-time curvature variant of the mass R a b=inertial mass of the black hole such that

$\mathbb{R}n = R\ abc \otimes 1\ where\ R\ abc\ constricts\ with\ near\ infinte\ curvature\ from R\ g\ ab$

and R ab is a huge number .The expression$\prod \frac{1}{2n\pi} \rightarrow \frac{1}{n\pi}\ \ g\ ab = \frac{1}{\infty} = 0\ space -$

$time\ with\ near\ infinite\ curvature\ in\ a\ black\ hole\ with\ constricted\ space. Here\ \mathbb{R}n \rightarrow$

$0\ as\ \ n\ approaches\ a\ large\ number. The\ expression\ \rho\ ab\ relates\ to\ R\ ab\ as\ the\ enrgy\ equivalent$

of the inertial mass R a b and incorporates the 1/c^2 or 1/p^2c2+m^2^c^4=energy and p=momentum incorporates into

$\rho\ ab\ the\ energy\ density\ with\ regard\ to\ the\ metric\ g\ ab.$Of course the

1/2

$\pi\ relates\ to\ the\ spherical\ nature\ and\ circumference 2\pi\rho\ at\ the\ event\ horizon\ of\ a\ black\ hole\ wl$

space-time is constricted down toward 0 with infinite curvature and as the areas where the energy density of

matter

$\rho\ and\ inertial\ mass\ constrict\ more\ and\ more\ it\ goes\ from \frac{1}{2\pi} to \frac{1}{4\pi} to \frac{1}{8\pi} times\ the\ metric\ g\ ab\ an$

stress energy T

ab$\rightarrow$ $1. as\ inertia \leftarrow gravity. As\ \rho\ ab \rightarrow$

$large\ number\ the\ entire\ expression\ for\ \mathbb{R}n \rightarrow$

$0\ \ but\ where\ does\ the\ energy\ go\ from\ the\ event\ horizon\ of\ a\ black\ hole. Answer\ quasars\ with$

Hawking Radiation being spumed out like a jet engine leaving the black hole cold as a C02 cartridge would be cold after the contents of the cartridge were suddenly forced out

.Therefore

$\rho\ ab\ would\ manifest\ the\ energy\ for\ the\ inertial\ mass\ R\ ab\ like\ a\ C02\ cartidge\ being\ discharge$

over a protracted time period.

The expression

S=2

$\pi(NQ1Q5)^{1/2}$*describes black hole entropy in terms of it's* 252 *different disparate states* .This would be similar in some ways to $2\pi\rho = circumference(space - time) \rightarrow$ 0 *so* $\rho \rightarrow 0$ *deep within a black hole* and the diminishing sphere of space-time to a point is described by $\mathbb{R}n = \frac{1}{2^{n\pi}as} n \rightarrow \infty\ gab \otimes Rabc - 0.5Rg\ ab \otimes \rho\ ab\wedge - 1$ where space-time=$2\pi\rho$.

Note that a circle reducing to a point is a cone or asymptotically a spiral. Therefore the operator $\Pi n = 1\ to \infty$ ½^nπ would be the spiral operator on space-time reducing it from a sphere to a point in a cone shape The Spiral operator would be $\Pi \frac{1}{2^n}\ \pi^{-1}$*where the infinite product* Π *is from n to* ∞ eigen-states.

The Bekenstein Hawking Equation for black hole entropy

S=A/4Lp^2c^3A/4G$\hbar$ *where* $\hbar =$ *Planck'sConstant*$6.63x10^{-34 joule}$ – sec *or meters*^{2kg}*/sec G =* *gravitational constant*$6.67x10^{-11}$*newton meters*/sec ^2 *Lp = Planck length =* $10\wedge - 33cm$ A=cross sectional area or kA/Lp^2 where k=Boltzman constant and

Pl=$\hbar \frac{G}{c^3} =$

10^{-33cm}*again. Black hole entropy(S) based on this is* $\frac{1}{4}$ *or also Steven Hawking postulated*

0.29.Black hole entropy is directly proportional to the area of the event horizon by the Boltzman Constant k which was explained earlier with the Boltzman Equation for different states of matter.This is the maximum entropy obtained by what's called the Berkenstein Bound and relates to the Holographic Universe and principle of a two dimensional fingerprint of information in the black hole at the event horizon. Supersymmetry was applied to black holes using D-branes and string theory duality with regard to the SO(32)string theory and the compactification of closed string theory IIa to a circle with regard to M theory.

The zeroth law states the surface gravity of the event horizon of a stationary black hole doesn't vary. The first law states that the the change in energy dE=k/8$\pi dA + \Omega dJ + \Phi dQ$ *where* $\kappa =$ *surface gravity A = area of event horizon*$\Omega =$ *angular velocity J = angular momentum*$\Phi =$ *electrostatic potential of the charge Q* The second law is that the horizon area is a non- decreasing function with regard to time or dA/dt> or =0.Whenit was discovered that Hawking Radiation was emitted by black holes and the area and mass(therefore gravity) decreased over time this was became a" weak law". The third law of black holes is that

k

or κ *for surface gravity cannot be* 0 *The zeroth law states that surface gravity is similar* to thermal equilibrium in thermodynamic systems in that a temperature doesn't vary neither does surface gravity κ*in a black hole*.Wikipedia;Bekenstein Entropy

In the equation S(BH)=$2\pi\sqrt{NQ1Q5}$ N is the number of states as mentioned before and Q1 relates to the one-brane which relates to effecting or carrying the electrical charge related to the monopole or the electron emanating from higher states of matter in terms of n-branes.Q5 relates to the charge relating to space-time as the electron has a cloud with a probability density that goes out to ∞ *without reaching it but bounded by the event horizon of a a black hole.*

Based on this S=$2\pi\sqrt{(252)}$*charge of an electron gas(space − time)*

Charge of an electron is 1.6x10^-19 coulombs and the 5brane relates to near infinite curvature of a contracted area of space-time as told by
$\frac{\Pi 1}{2n\pi}$ $g\ ab\ \ R\ abc - \frac{1}{2} r\ g\ ab \otimes\ \rho\ ab^{-1} where\ n =$
number of eigenstates in n dimensional space. for the infinite product from n to∞ giving spa

½^nπ $as \frac{1}{2^{0}\pi} as\ space - time\ approaches\ the\ D - 0\ state\ or\ D - 0 -$
branes making $\frac{1}{2\pi} = \frac{1}{2\left(\frac{22}{7}\right)} = \frac{7}{22(2)} = \frac{14}{22}$ *so* 44/

7$\sqrt{252(1.6x10^{-19})(\frac{14}{22})}$=6.14[(160.36)(1.6x10-19)]^1/2 which is 6.141(6x10-17=36.6x10^-17 as black hole entropy in the zero(0)dimensional state or the D-0-brane which is very close to 0 entropy while Steven Hawking postulated entropy of a black hole to be approximately 0.29. This is because while spacetime is constricted at the event horizon it still exists so there is no 0 dimensional state within a black hole which would make the 5-brane relate to a four or perhaps higher dimensional state with the 252 different states of matter including energy-matter conversion or entanglement due to the superhigh pressures and super cold temperatures. It is likely that Boso- Einsteinian Condensate would exist with a matrix that traps photons, electrons, positrons ,neutrinos ,anti-neutrinos ,bosons and at near the zero dimensional state fermions. It is also possible that tachyons would be trapped in a black hole which would reverse time's arrow and have a negative mass or strange mass hybrid between negative and positive mass.It is these tachyons that would cause the –mass in the superimposed region 3 onto region 2 that would solve the Hawking Paradox. Whether radiation actually takes on a liquid state is agnostic but not impossible under those conditions.
If the spiral operator operates on the function R a b c-1/2Rgab$\otimes$ $\rho ab^{\wedge} - 1$ and is inclusive of the metric g ab over the infinite product of eigenstates over n dimensional space it is asymptotic to Schwarzchild Space-time if the expression is reflected at the event horizon to express the increasing cone or spiral from approximately zero(0)dimensional space-time to n dimensional space-time where n goes from the D-0 eigenstate to D-n-eigenstate.This can be accomplished if
$\prod from\ n = 11\ or\ n = 10\ decends\ to\ n =$
$0\ for\ the\ \ D -$
brane which gives an opening spiral or cone past the event horizon as there is a descending cc

Cone at the event horizon with space-time spiraling in from ordinary space.The factor relating to the mass equivalent on both sides of the event horizon emanates from the function on which the operator is operating or "The Equation of Everything" so Schwarzchild

spacetime=$\prod n \frac{1}{2^{n\pi}} -$

$\frac{\Pi n1}{2^{n\pi}g} ab$ *where the limits of the infinite product*Π *are from* n *to* $\rightarrow$

D 11? *and from*$D11 \rightarrow n$ *where* n *starts at* $D = 1$ *not* $D =$

0 *as the spiral is a modified cone acted upon by the metric* $g\ ab$ *as per* $Rabc -$

$\frac{1}{2R} g\ ab \otimes (\rho\ ab)\text{^} - 1.$

CHAPTER NINTEEN
DARK ENERGY AND DARK MATTER

It was questionable about whether dark energy and dark matter were related until the missing mass calculation for the mass equivalent of dark energy and the mass of dark matter were found to be congruous. As so they share charcaterisitcs and can be considered in the same gauge symmetry group.This can be negociated with the mechanism of the "Big Bang" which is like lighting a match. From a chemical standpoint(as previously mentioned)the unlit phosphors would be the homogeneous antimatter-matter mix,the Big Bang would be the light being struck,and the charred residual would be the dark matter while the energy expended would be the Dark Energy coupled with the energy from the annihilation of matter and antimatter in the "Big Bang".

As mentioned previously the antiparticle-antiparticle repulsion o\in the quantum bubble under extreme pressure and temperature force a huge anti-gravitation force to push matter with space-time outward after the rotational component of "The Big Bang" had almost completely slowed toward 0. The anti-gravitational force pushing galaxies apart from each other is Dark Energy and the residual antimatter that was burned out is dark matter.As the anti-gravitational moment of the interaction was carried off by the Dark Energy ,Dark Matter has a gravitational effect instead of antigravity although it emanated from anti-matter.

As mentioned previously the particles of dark matter are on and about everything just as the BMR from the "Big Bang" is all pervasive ,but the particles of dark matter are so small yet homogeneous that they might be a multiple of

Pl=

$\hbar \frac{G}{c^3 or} 10^{-33}$*cm which would make the density of this superfine powder or dust have a very sli*

slight measurable density in the massive volume of space although the gravitational effects from the mass are considerable although measured indirectly. This too was already mentioned.

It has been postulated that Dark Matter is composed of baryonic particles and neutrinos possibly with anti-neutrinos and anti-Hadrons as a residual from anti-matter and that Dark Matter acts as a type of "cosmic glue" which has been present since the "Big Bang"

There is a question as to how much of the cosmologic constant
Λ *is related to Dark Energy, but the force to push the* 750 *billion galaxies apart from each oth*
seems to be far greater than what was empirically measured for Dark Energy and greater thanΛ(*the cosmologic constant*)*which relates to* $8\pi G/c$^4.

With a stretch of the imagination and some creative math one can see that the 1st,3rd,5th and all odd dimensions have dark matter sequestered as it's gravitational effect can only be indirectly measured and mass in the even dimensions 2nd,4th,6th,8th etc. can have it's properties directly measured .This can evolve from the space-time formula of-1/2e^-i n cot
θ *where* $\theta = \pi$ *radians which is the trajectory of the "Big Bang"*.Odd powers of i give results containing I while even powers of I give real numbers such as i^2=-1 ,i^4=-1 etc while i^1=I and i^3=-I and i^5=-I such that if this was the scenario due to the inert nature of dark matter its mass may be postulated mathematically mass of dark matter=mass of ordinary matter/i *or*$\sqrt{-1}$.This would be likely if dark matter which is anti-matter based showed anti-gravitational rather than gravitational effect as Fg=Gm1m2/r^2 where 6.67x10-11n-m/sec^2 or 10-11=G and m1=mass/i m2=mass2/i such that Fg=Gm1m2/i^2r^2=-1(G m1m2/r2)which would indicate mutual repulsion of antiparticles and anti-gravity .While it may be true that the odd dimensions may have dark matter and possibly most of the missing mass equivalent of Dark Energy forming the difference between what is measured and mathematically calculated as a prediction on we will still have to show strange or oscillating mass in dark matter(as burned out antimatter).This oscillating property would be between + and – mass and while experiments with the Hadron Collider in Cern, Switzerland are looking for anti-gravity between anti-particles and particle-anti-particle interaction before they annihilate ,it has not clearly been demonstrated yet that anti-particles display pure anti-gravity although they do display some properties of anti-gravity.It also has not been demonstrated that anti-particles display anything except a positive mass.The question is this .How does experimentally show burned out antimatter in dark matter when it's effects can only be indirectly measured. Also is mass/i the same as strange or oscillating matter with a hybrid between +mass and –mass. If this property follows antimatter and antiparticles it may also follow dark matter with much huger antigravity with Dark Energy. Mathematically shadow odd dimensions cannot be ruled out but whether or not they contain the mass of dark matter or mass equivalent of dark energy is up to speculation.

CHAPTER 19

IS THE FABRIC OF SPACE COMPRISED OF STRINGS?

The fabric of space is comprised of virtually massless leptons and gluons holding the fabric together. Space existed before the FIRST EVENT DURING THE TIME OSCILLATION PARADOX WHWEW SPACE-TIME DEVELOPED A SPINS AND CENTRIFUGE EFFECT FORMING A VORTEX IN WHICH THE MULTI-VERSE OR AT LEAST OUR UNIVERSE WAS AT THE CENTER . STRINGS EXISTED AFTER THE FIRST EVENT WHEN GRAVITY FORMED WITH THE SPIN 2 VECTOR BOSONS EMERGING FROM THE FERMIONIC OR VACUUM STATE FROM WHICH THE FERMIONS OF LEPTONS AND GLUONS FORMED BOSONS AND QUARKS) which entail the STRONG FORCE)and are comprised of gluons which always existed. STRINGS HAVE A MASS AND AS A RESULT ARE CONSIDERED MATTER OR ENERGY; NATURALLY THIS ENERGY IS POTENTIAL AND KNIETIC ENERGY; yet the cosmologic constant which repelled the infinite parallel planes preventing them from forming dimensions by their touching always existed as the energy density of a vacuum is the cosmologic constant. As a consequence the cosmologic constant always existed as did space. If the components of space are truly massless then it is not necessarily comprised of strings and they formed after the first event, but if leptons and gluons have a miniscule mass almost immeasurable(as in the case of photons) then they could indeed be comprised of strings and strings could have always existed. If strings are composed of pure energy then they existed before the first event but if they are compromised of matter and energy or pure matter then they existed only after the first event. As strings have a miniscule mass at least part of the string is composed of matter but photons also have miniscule mass and photons are electromagnetic radiation so energy must at least partly composed of matter. Even space with potential energy have miniscule kinetic energy so there is an asymptotic value of matter involved in leptons and gluons which comprise and hold together space. The cosmologic constant reflecting antigravity always existed and was caused by the miniscule mass of space but if gravity existed before the first event it is so small that it approaches zero without reaching it, Based on this one must compare the miniscule mass of each string with the miniscule mass of leptons ,gluons and photons. If the mass of strings is greater than that of leptons or gluons and possibly photons then STRINGS OCCURRED AFTER THE FIRST EVENT ONLY,but if less THEN STRINGS ALWAYS OCCURRED AND WHERE EXISTANT BEFORE THE FIRST EVENT.

CHAPTER NINTEEN CONTINUED

DARK MATTER AND DARK ENERGY

Using Ockham's(Occum's)Razor which scenario is most likely in our universe with it's laws for our immediate reference frame? When unaccounted for mass is indirectly measured by gravity measurements and there is no other clear objective measurable evidence of it's existence is dark matter in odd dimensions of space-time possibly showing a parallel universe which has an opposite rotational vector in space-time(The Anti-swirl to the swirl of the rotational vectors of "The Big Bang" where space-time=+1/2e-i n cotθ or -1/2 e^in cot

θ *where the sum total of all space* –

time curvature is 1 *or asymptotic flatness or the dark matter and dark energy are biproducts*

of "The Big Bang" where Dark Matter is the residual of the anti-matter and Dark Energy was the anti-gravity. Also to be added that the micro-particles or micro-dust of dark matter is so small and homogeneous that it's mass is hidden by the enormous(in comparison)volume of space. Obviously there is very little evidence of parallel universes but it can't be ruled out. However with our level of technology it should be provable that micro-particles of dark matter do exist especially with the Hadron Collider .How does one prove the strong anti-gravitational effects of Dark Energy when only weak readings are being recorded. Obviously something must be getting missed in the measurements although reproducibility implies precision ,there may be some soft?f actors that are being underweiged or over weighed. The comes the third most sobering possibility which is probably unlikely ,that dark matter readings from gravitational perturbations and the missing mass is due to measurement or reading errors due to some soft factors(Compton effect?)and the rate of the recession of galaxies from the Red Shifts might be off because of error in the speed of light due to incorrect readings of photonic mass. This author does not want to think that the third scenario is true though it can't be totally ruled out even twenty years or more since the hypothesis of dark matter was first postulated. Ruling out scenario #3 it seems most likely that dark matter is a byproduct of" The Big Bang"just as the BMR is and that there is a micro-dust adherent to virtually all mass as a burned out residue from antimatter while Dark Energy is the anti-gravitation force from the antimatter "burn out". Still, it is possible that odd dimensions in space-time reflect a parallel manifold just as CW manifold would be complemented by a CCW manifold and that dark energy and dark matter have most of the cause of the measured effects in those odd dimensions. Then there is scenario 4 which is that scenario's 1,2 and 3 are all wrong partially wrong or inaccurate. The scenario of the fine microdust also explains the mass

gap$\Delta t >$

0 *in Yang Mills Theory with another non* –

Abelian gauge symmetry group with Dark energy and Dark Matter. This scenario would close

the mystery of
the$\Omega -$
particle in the gauge symmetry group $SU(3)$*involving the Strong Force.*

Does the change of the existence of Dark Matter change with the presence or absence of an intelligent observer .In other words can "Spooky Action at a Distance"or entanglement be applied in this case as with the disappearing and reappearing of an electron(which has a gas or region extending out to other electron gasses or regions at vast differences with their exchanging information). This particular author is agnostic(questioning the validity) about the idea that Dark Matter appears and disappears whether or not an intelligent observer exists taking measurements. Although Schrodinger's Cat implies that this scenario could occur where an observer could have dark matter shift into the odd dimensions from the even dimensions and revert to the even dimensions when no intelligent observer is present. There is mathematical data to support this as there is mathematical data on space-time to indicate the slippage of Dark Matter from visible measurable dimensions to invisible indirectly measurable dimensions there is a paucity of objective evidence to support this hypothesis and it still isn't "the simplest explanation of the phenomenon "as per Occum's Razor. Still are all things equal? If not Occum 's razor would not apply.Still "The Inter-connectiveness Theorem of Bohm"has an experiment shown on a Discovery Channel broadcast in the 1990's in which ink in a glass of water or solution swirled one way would made the ink disappear completely and swirled the other direction would made the ink reappear and start to dissolves .It was postulated that the ink was sliding to and from our measurable dimensions which can be observed to dimensions where measurement would have to be indirect and cannot be observed .Does this experiment support the idea that Dark Matter is slipping into odd dimensions which can only be indirectly observed and back to observable dimensions when there is no observer .If so there should be a counteraction to the action which made Dark Matter stay in the unobservable dimensions just as swirling the ink solution in the opposite direction makes the disappearing ink re-appear .If so what would be the counter-action ?Does the above make this scenario more likely ?The Inter-connectiveness Theorem states everything is connected or interconnected with everything else which would help explain "Spooky Action at a Distance". How valid this this assertion? The way this author sees it there isn't enough empirical data to either support or refute it and if there are alternative explanations to the swirling ink experiment that might refute it partially. However with Superstrings there may be a type of inter-connectiveness. within supersymmetry groups such as the SO(32)gauge symmetry group and overlapping gauge symmetry groups.

Dark Energy relates to a scalar trace of a field acting on a 5-brane with regard to a 4-brane asΦ $where\ \frac{d2\Phi}{dt^2} + 3\frac{Hd\Phi}{dt} + V(\Phi) = 0\ where\ H^2 = \frac{8\pi}{3m^2} \quad \rho\left(1 + \frac{\rho}{2\sigma}\right) + \frac{\Lambda 4}{3} +$
$\mathcal{E}\frac{}{a4}$*where the tension on the* $4 -$ *brane from the* $5 -$
brane is σ *and* ρ *is the energy density of matter where*$\mathcal{E}$ *is an integration constant where*

space-time curvature is acting on the 4-brane,The stress energy tensor relates to the Ricci Tensor via Einstein's Field Equation for Relativistic Gravity where

$8\frac{\pi G}{c^4} \quad Tab =$
$R\ ab -$
$\frac{1}{2} R\ g\ ab$ *where the stress energy on the brane relates to the Tension or* σ *on the brane.*

With regard to anti-gravity $8\pi\Lambda$ *can be substituted for* $- \ 8\pi G$ so
$8{-}8\pi G$ *becomes* $8\pi\Lambda$ *and* $\frac{8\pi\Lambda}{c^4} \quad Tab =$
$R\ ab - 1/2(-R\ g\ ab)$*to show reciprocal curvature for anti – gravity* $= Rab -$
$\left(-\frac{1}{2}\right) R\ g\ ab = R\ ab + \frac{1}{2R} g\ ab$ so -$8\frac{\pi G}{c^4} \quad T\ ab = R\ ab + \frac{1}{2} R\ g\ ab$=$8\frac{\pi\Lambda}{c^4} \quad T\ ab =$
$G\ ab = 0.$

CHAPTER TWENTY-ONE

SPOOKY ACTION AT A DISTANCE

The phenomenon of " Spooky Action at a Distance" relates to the exchange of information such as electron spin over great distances which may change when an observer is present. This phenomenon with regard to electrons may be caused by an electron cloud which doesn't dissipate over great distances meaning the fingerprint or signature of the electron's information could exist kilometers away or farther distance which might not seem possible with normal physics. A similar phenomenon can occur with photon pairing although photonic information with the wave particle duality can travel at just under "c" if photons have a non-resting mass and "c "if photons are massless. The pairing of information between photons could be because the wave property of photons could be analogous to the electron cloud of an electron and since photons are virtually ageless and immutable, information is also immutable must be preserved and can be exchanged between paired photons over huge distances via the wave property of photon via De Broglie's Equation$\lambda =$
$\hbar \frac{}{mc}$ *where masslessness* $m =$
0 *makes an almost infinite wavelength and super high frequency with very high energy*

From the Planck's Equation
E(photon)=$\hbar \upsilon$ *where* $\upsilon =$
frequency. Based on these equations very low frequency waves or photons such as radio wave.

have more of a mass than high frequency waves such as x or gamma rays. Massive objects also have a corresponding wavelength regardless of the velocity of the massive object ($\lambda = \hbar \frac{}{mv}$ *and the wavelength gets shorter as the velocity* $\rightarrow c$.
The observed positions of electrons was not where they were supposed to be due to electron cloud probability distributions of |.e E|>πm^2/ ln 2 where e=electric charge(Q) and E- electric field and mass is electron mass and probability of a generator penetrating a sphere of diameter of z such that |z| leads to an asymptotic divergence of probability with a continuity limit of varying field strength and the definition of probability being violated by the tail of the distribution .With regard to carrier densities, the density distribution of matter is
gc(E)=8
$\pi \sqrt{\frac{2}{\hbar^3} \quad \frac{me^3}{2}}$ *where* $\hbar =$ *Planck'sconstant and*$\sqrt{E - Ec}$ *for* $E > Ec$ *has Ec as the conductioon bar*

between the density of energy states and is where r=0 on the sphere upon which the generating function acts where 2r=z. In this instance "g" is the scalar metric not G .Discrete levels of Normalization for electrons in the volume of an infinite diameter sphere can be summated into a convergent system where the field strength and asymptotic behavior of the probability distribution acts on all particle masses with carrier densities as the infinite diameter of the sphere is forced to converge at the

circumference
or $\pi|z|^2 where\ z = diameter\ and\ |z|^2 is\ the\ probability\ distribution.$ Σz=0 to
z→ ∞$(\frac{4}{3\pi r^3})\hbar(E - Ec)$where the volume of the sphere is
$4/3\pi r^3 and\ \ z\ is\ the\ diameter\ with\ the\ infinite\ sum\ of\ the\ \ system.$

Based on the wave function with Schrodinger's Equation the probability distribution of the electron field density with regard to the charge of an electron will diverge to $\infty\ as\ \ z \rightarrow$
$\infty\ until\ the\ volume\ is\ such\ that\ z\ converges\ with\ the\ circumference\ which\ must\ be\ at\ time =$
$\infty\ according\ to\ relativity\ as\ the\ circumference\ is\ expanding\ faster\ than\ the\ diameter(z).$

As the electron cloud distributions of many or all electrons in a field which can stretch out vast distances information on spin can be exchanged between electrons in the electron gasses that compose the field.

The Schrodinger Equation$\psi(E) = gc(E) = \frac{8\pi\left(\frac{2me^3}{\frac{2}{\hbar^3}}\right)^1}{2}\ \ \sqrt{E - Ec\int_{Ec}^{\infty} gc(E)F(E)dE}$

Where n= carrier density f (E)=Fermi Dirac Proabability Function and
n0=$\int_{Ec}^{\infty}\frac{8\pi\left(\frac{2}{\hbar^3}\right)^1}{2} or\ 8\pi\sqrt{2/\hbar\verb|^|3}$ me^3/2$\sqrt{Ev - E^1 + ef}$ E/kT dE where
k=Boltzmann constant exists for each specific state of an electron gas. The action(S) of the metric g for an electron with a carrier density of E v acting on the electron gas has a 100% probability with localization of the electron gas at infinite diameter z in the expanding sphere of space-time.The action S=c^8/12$\frac{\pi GR1R2(-EvE)^3}{2r} where\ R1 =$
$p0\ and\ R2 =$
$n0\ and\ P0\ is\ the\ probability\ of\ the\ carrier\ density\ of\ E\ v\ as\ a\ state\ of\ matter =$
$\rho 0. With\ R1\ \ as\ carrier\ density\ of\ state\ of\ matter\ acting\ on\ n0\ of\ the\ electron\ the\ action$

Of the carrier density n has a probability of 100 % for the electron gas acting on
$\rho\ the\ energy\ density\ of\ all\ states\ of\ matter\ as\ determined\ by\ the\ Boltzman\ Equation\ where$

K=Boltzman constant and T=temperature in degrees kelvin. Therefore there is a 100% probability that the carrier density of an electron gas will interact with the density of matter in a diameter(z) traveling an infinite volume and not converging with the circumference until the diameter and the circumference meet which is why electron gas paths must intersect and exchange information such as spin.

With regard to photons which are purported to be massless and traveling at "c" while electrons travel at approximately v=0.65 c photon pairing is much simpler to explain in terms of contracted space and dilated time. Information of a photon(as dictated by the energy level=$\hbar v$) *travels vast distances at "c"* and the information may actually also change with or without an observer assuming that the temperature is 2.74 degrees kelvin or higher .To travel faster than this the photon would have to be a wave or electromagnetic radiation rather than a particle bending or warping contracted space to transmit the information. Photons and waves are supposed to travel in a double helix path burrowing spirals into space rather than in a strictly straight line(if photons were actually massless)Despite this the "Spooky Action(S)at a Distance"is still quite spooky and resembles the technique for a transporter device.

CONCLUDING REMARKS

The purpose of this book is to introduce to the scientific community mathematical adaptations of concepts regarding the five duel string theories and M Theory in terms of "The Equation of Everything" and applying this equation to Black Holes with the Spiral Operator to demonstrate it's congruence to Schwarzchild Spacetime.In addition this author has attempted to solve the "Hawking Paradox"using superimposed regions of Schwarzchild Space-time with negative mass from tachyons(replacing + mass bosons) to net out the information over dilated time as 0 so information isn't lost in a Black Hole. Also,the existence of Dark Matter has perplexed physicists for 25 years and in this work this author has attempted a new theory based on math that Dark Matter exists on everything as a fine microscopic powder with significant mass but in the volume of the universe has no measurable density making it invisible. In addition there was a purported mechanism that when anti-matter and matter either blew out with "The Big Bang" at the 0,0 point and inflated out with Inflation.Again consider light a match with the light being the Big Bang,the residue on the top of the match is Dark Matter,the force of anti-gravitational(push)is Dark Energy and the phosphors of the unlit match is the anti-matter matter mix.The residual matter after the anti-antimatter repulsion and anti-matter matter annihilation is the dark matter and ordinary matter after "The Big Bang".

The fact that string theory can be compactified(curled up)into a circle as IIa string theory and possibly M Theory lead one to realize that the equation of arc length s=rθ *can relate to spacetime and the Riemann Forces such that space − time accelerates and the Riemann forces curve spacetime as increments of* θ *from* 2π *radians*

2n

π *radians where the circumference of the circle(compactified IIa closed string theory and M t*

theory outruns the diameter of the circle which represent the Riemann metric in both directions with $\mathbb{R}ab = \mathbb{R}^\wedge ab$ such that the radius is the + or – Riemann metric. In this case as the circumference of space-time is

$2\pi R$ *and since* $2R = d$ *then the rate at which space − time is" outrunning"* the Riemann forces is

πc *as space −*

time and perhaps tachyons are the only quantities that can exceed the speed of light boundar)

CHAPTER TWENTY-ONE

SPOOKY ACTION AT A DISTANCE

The phenomenon of " Spooky Action at a Distance" relates to the exchange of information such as electron spin over great distances which may change when an observer is present. This phenomenon with regard to electrons may be caused by an electron cloud which doesn't dissipate over great distances meaning the fingerprint or signature of the electron's information could exist kilometers away or farther distance which might not seem possible with normal physics. A similar phenomenon can occur with photon pairing although photonic information with the wave particle duality can travel at just under "c" if photons have a non-resting mass and "c "if photons are massless. The pairing of information between photons could be because the wave property of photons could be analogous to the electron cloud of an electron and since photons are virtually ageless and immutable, information is also immutable must be preserved and can be exchanged between paired photons over huge distances via the wave property of photon via De Broglie's Equation$\lambda =$
$\hbar \frac{}{mc}$ *where masslessness* $m =$
0 *makes an almost infinite wavelength and super high frequency with very high energy*

From the Planck's Equation
E(photon)=$\hbar v$ *where* $v =$
frequency. Based on these equations very low frequency waves or photons such as radio wave.

have more of a mass than high frequency waves such as x or gamma rays. Massive objects also have a corresponding wavelength regardless of the velocity of the massive object ($\lambda = \hbar \frac{}{mv}$ *and the wavelength gets shorter as the velocity* $\rightarrow c$.
The observed positions of electrons was not where they were supposed to be due to electron cloud probability distributions of |.e E|>πm^2/ ln 2 where e=electric charge(Q) and E- electric field and mass is electron mass and probability of a generator penetrating a sphere of diameter of z such that |z| leads to an asymptotic divergence of probability with a continuity limit of varying field strength and the definition of probability being violated by the tail of the distribution .With regard to carrier densities, the density distribution of matter is
gc(E)=8
$\pi \sqrt{\frac{2}{\hbar^3} \quad \frac{me^3}{2}}$ *where* $\hbar =$ *Planck'sconstant and*$\sqrt{E - Ec}$*for* $E > Ec$ *has Ec as the conductioon bar*

between the density of energy states and is where r=0 on the sphere upon which the generating function acts where 2r=z. In this instance "g" is the scalar metric not G .Discrete levels of Normalization for electrons in the volume of an infinite diameter sphere can be summated into a convergent system where the field strength and asymptotic behavior of the probability distribution acts on all particle masses with carrier densities as the infinite diameter of the sphere is forced to converge at the circumference

or $\pi|z|^2$ *where* $z =$ *diameter and* $|z|^2$ *is the probability distribution.* Σz=0 to $z\rightarrow\infty(\frac{4}{3\pi r^3})\hbar(E - Ec)$ where the volume of the sphere is $4/3\pi r^3$ *and* z *is the diameter with the infinite sum of the system.*

Based on the wave function with Schrodinger's Equation the probability distribution of the electron field density with regard to the charge of an electron will diverge to ∞ *as* $z\rightarrow\infty$ *until the volume is such that z converges with the circumference which must be at time* $= \infty$ *according to relativity as the circumference is expanding faster than the diameter(z).*

As the electron cloud distributions of many or all electrons in a field which can stretch out vast distances information on spin can be exchanged between electrons in the electron gasses that compose the field.

The Schrodinger Equation $\psi(E) = gc(E) = \frac{8\pi\left(\frac{2me^3}{\frac{2}{\hbar^3}}\right)^1}{2}\sqrt{E - Ec\int_{Ec}^{\infty} gc(E)F(E)dE}$

Where n= carrier density f (E)=Fermi Dirac Proabability Function and

n0= $\int_{Ec}^{\infty}\frac{8\pi\left(\frac{2}{\hbar^3}\right)^1}{2}$ *or* $8\pi\sqrt{2/\hbar\text{^}3}$ me^3/2 $\sqrt{Ev - E^1 + ef}$ E/kT dE where k=Boltzmann constant exists for each specific state of an electron gas. The action(S) of the metric g for an electron with a carrier density of E v acting on the electron gas has a 100% probability with localization of the electron gas at infinite diameter z in the expanding sphere of space-time.The action S=c^8/12 $\frac{\pi GR1R2(-EvE)^3}{2r}$ *where* $R1 = p0$ *and* $R2 = n0$ *and* $P0$ *is the probability of the carrier density of* $E\ v$ *as a state of matter* $= \rho 0$*. With* $R1$ *as carrier density of state of matter acting on* $n0$ *of the electron the action*

Of the carrier density n has a probability of 100 % for the electron gas acting on ρ *the energy density of all states of matter as determined by the Boltzman Equation where*

K=Boltzman constant and T=temperature in degrees kelvin. Therefore there is a 100% probability that the carrier density of an electron gas will interact with the density of matter in a diameter(z) traveling an infinite volume and not converging with the circumference until the diameter and the circumference meet which is why electron gas paths must intersect and exchange information such as spin.

With regard to photons which are purported to be massless and traveling at "c" while electrons travel at approximately v=0.65 c photon pairing is much simpler to explain in terms of contracted space and dilated time. Information of a photon(as dictated by the energy level=$\hbar v$) *travels vast distances at "c"* and the information may actually also change with or without an observer assuming that the temperature is 2.74 degrees kelvin or higher .To travel faster than this the photon would have to be a wave or electromagnetic radiation rather than a particle bending or warping contracted space to transmit the information. Photons and waves are supposed to travel in a double helix path burrowing spirals into space rather than in a strictly straight line(if photons were actually massless)Despite this the "Spooky Action(S)at a Distance"is still quite spooky and resembles the technique for a transporter device.

CONCLUDING REMARKS

The purpose of this book is to introduce to the scientific community mathematical adaptations of concepts regarding the five duel string theories and M Theory in terms of "The Equation of Everything" and applying this equation to Black Holes with the Spiral Operator to demonstrate it's congruence to Schwarzchild Spacetime.In addition this author has attempted to solve the "Hawking Paradox"using superimposed regions of Schwarzchild Space-time with negative mass from tachyons(replacing + mass bosons) to net out the information over dilated time as 0 so information isn't lost in a Black Hole. Also,the existence of Dark Matter has perplexed physicists for 25 years and in this work this author has attempted a new theory based on math that Dark Matter exists on everything as a fine microscopic powder with significant mass but in the volume of the universe has no measurable density making it invisible. In addition there was a purported mechanism that when anti-matter and matter either blew out with "The Big Bang" at the 0,0 point and inflated out with Inflation.Again consider light a match with the light being the Big Bang,the residue on the top of the match is Dark Matter,the force of anti-gravitational(push)is Dark Energy and the phosphors of the unlit match is the anti-matter matter mix.The residual matter after the anti-antimatter repulsion and anti-matter matter annihilation is the dark matter and ordinary matter after "The Big Bang".

The fact that string theory can be compactified(curled up)into a circle as IIa string theory and possibly M Theory lead one to realize that the equation of arc length s=rθ *can relate to spacetime and the Riemann Forces such that space − time accelerates and the Riemann forces curve spacetime as increments of* θ *from* 2π *radians*

2n

π *radians where the circumference of the circle(compactified IIa closed string theory and M* ı

theory outruns the diameter of the circle which represent the Riemann metric in both directions with $\mathbb{R}ab = \mathbb{R}\text{^}ab$ such that the radius is the + or – Riemann metric. In this case as the circumference of space-time is

$2\pi R$ *and since* $2R = d$ *then the rate at which space − time is" outrunning"* the Riemann forces is

πc *as space −*

*time and perhaps tachyons are the only quantities that can exceed the speed of light boundar*ȝ

CHAPTER 22

IS THIS UNIVERSE TWO DIMENSIONAL?

Flat matter consists of photons ,strings(open and closed)and anything else perceived as a holographic projection. The 2 dimensional lattice equation of Michio Kaku indicates a relationshipship between quantum mechanics and relativity with regard to the critical exponent when everything comes apart. Could this relate to "The Big Bang" or "Big Crunch'?
The curling up or compctification of string theory or M Theory is a circle not a sphere and the dimensions which are greater than two in a holographic construct of the world sheet are compactified dimensions and string sized forming membranes.

White noise utilizing CMB data contains trillions of bits of information which are string sized 10^-33 cm or smaller In this author's first book "Megaphysics,A New Look at the Universe"(2003)it was postulated that a two dimensional universe would exist as would a world sheet in string theory helping to unify quantum mechanics and relativity. This would include space in dilated time comprised of almost massless photons of electromagnetic radiation which would travel at all speeds except zero depending on the medium they pass through and because influenced by gravity such as in black holes they do have miniscule mass as does space. So in essence the vacuum or fermionic state of matter may be comprised of photons with gluons holding space together ,while tachyons still exist above the velocity of photons traveling backwards in time which would be infinitely slow if time were totally dilated. The event horizon or any active black hole contains bits of information which would or might appear to the observer as being two dimensional as well as in dilated time and photons are postulated to have existed virtually always. In addition strings whether open or closed are comprised of flat matter or energy which is also flat matter and therefore two dimensiona lThe critical exponent of the 2D lattice equation may relate to the Big Bang w ith reference to criticality as might a Big Crunch.

'The Holographic Universe 'by Dr. Micheal Talbot indicates that two dimensional projections would exist in terms of the World Sheet as purported by string theory as strings are two dimensional. Paranormal phenomena like "ghost sitings"could be two dimensional images on a background three dimensional plane as "ghosys" are indicated as "white noise" or in the low energy fradio wave band as electromagnetic raidiation mwhich is comprised of photons which are flat matter or two dimensional.

A supposedly absurd theory purported that our universe is a computer program supported by physicisits like Dr. Neil deGrasse Tyson of the Hayden Planetarium may seem absurd but if this uniberse is comprised of flat matter which has two macroscopic dimensions and a near infinite number of string sized dimensions which comprise a super-brane or supermembrane as purported in M Theory then this would be compatible with a computer program comprised of a multi-googlplex of bits of data or information forming the program.This is compatible with the dolution to the Hawking Paradox which shows that the bits of information at the evetnt horizon of any active black hole is dispersed or dissolved in space-time as a diffusive process such as a suspension where the bits are so miniscule that they can't be detected by any present technology. What would control the program though?The Higgs Field comprised of tachyons and the massive Higgs Boson and how would the "computer" be turned on and off? Is the Big Bang where a rogue tachyon drops to the speed of light causing a a time oscillation paradox and the vortex of space-time from an infinite number of parallel planes be the turning on by the deterministic rogue tachyon dropping from above light speed to the speed of light causing the time oscillation paradox.And would a Big Crunch be the turning off of the computer program? Many of these ideas would br very difficult to prove but new evidence just released.On the Jpurnal of Physics Review Letters there is evidence that a two dimensional matrix exists with googolplex of microdata which this author believes forms a super-BRANE. It solves problem's with Einstein's Theory of Gravvity as in the two dimensional state the curvature of space-time caused by mass reaches as asymptototic limit approaching zero without reaching it in the two macroscopic dimensions while the superBrane emcompasses the other infinite dimensions.This two dimensional world sheet helps to unify quantum mechanics and trlativity and while the "Computer Program Hypothesis"seems unlikely and ridiculous;it isn't impossible if one can define what the computer ismthe program is and the operator of the computer is..Still it is the opinion of this author that this theory is based on out new technology based on computers and computer software and would have been unheard of in the 1970's or 80's and would have been laughed our of the scientific profession so many Physicisits would still like askance at this theory.

Chapter 22(continued)

THE TWO DIMENSIONAL UNIVERSE

THE TWO DIMENSIONAL MODEL BASED ON THE YANG BAXTER RELATION IS BASED ON A PAARTITION FUNCTION OF A TWO DIEMSNIONAL LATTICE FOR WHICH ELLIPTICAL JOSEPHSONVORTICIES ACT AS FLAT MATTER WITHIN A VIBRATING MATRIX OF PHOTONS TRAPPED WITHIN BOSOEINSTEINIAN CONDENSATE AT JUST ABOVE ZERO DEGREES KELVIN.THE EQUATION Z=$n\{\exp[-E(n)/kT$ where E(n)is the energy of the Nth state of matter,K=Boltzman Constant and T=temperature in degrees kelvin.The Free Energy(F)=-kT ln Z and Z is defined abpve.The Boltzman equation relates to energy states at phase transitions as in the latent state of freezing or mLf and of vaporization as mLv.At absolute zero there is a confluence of vibrating photons trapped in a matrix accounting for the mass gap in Yang Mills Theory and as mentioned previously in this book.The correlation functions between spins or isospins can be lbelled

σi andσj revealing criticality seeking behavior as the metric g ij regarding parameters of crit initial and j the final event.This expreession is $g\ ij = <\sigma i \sigma j> - <\sigma i><\sigma j>$ *and depends on the distance separating the states explained by the variable* xwith *regard to tl*

Distance and explains “Spooky Action at a Distance”as x approaches infinity as a distance between the states.g ij`x^- $\frac{\tau e^{-x}}{\zeta}$ *where* $\zeta =$ *correlation length and* τ *relates to relative time as* x *is the distance between states. As the corr*

Elation length approaches infinity we have a phase transition ind in any phase of flat matter such asElliptical Josephson Vorticies the temperature of the system can vary as there are 252 discreet states of matter in any active black hole. These states under superpressures included radiation which is flat matter in states such as liquid,solid BosoEinsteinian Condensate and would neatly fit as Elliptical Josephson Vorticies fitting into the vortex of space-time as it constricts toward zero as the event horizon of any active black hole is approached. This is one reason why a hologram or flat matter is imprinted at the event horizon in terms of microdata and evaporates or dissolves into space-time as microdate over a 360 degree area as well as via Hawking Radiation spuming outward from the event horizon.Basically these data are super-compactified into a two dimensional state where the other dimensions are super-compactified into a Super-brane while the two macroscopic dimensions can be viewed like a stop action photograph.The Ising Model has an energy operator as$\epsilon n = \sigma n \sigma n + 1$ *at criticality and reveal conformal invariance whereby the relation to conformral*

Field theory has a !:1 association with the fields of the Ising Model.

EXTREMOPHILES CAN LIVE IN THE ATMOSPHERE AND MOLTEN LAVA OF VENUS

Bacteria can adapt to extremes of temperature and radiation as mold has been found in high gamma ray sources and the Hadron Collider at Cern ,Switzerland.It is believed that after a cataclysmic planetary collision approximately 5 billion years ago whereby our moon was formed the atmosphere was extremely similar to that of Venus today. No water ,mostly carbon dioxide and nitrogen from which extremophic life formed in volcanic lava with temperatures similar to Venus today. This protolife resulted from a millisecond gamma ray burst from the fusion of two stars parsecs away which turned D.NA. and R.N.A in from inert organic compounds into chromosomes which are alive .Bacterial speciies were found to be able to withstand temperatures of 800 degrees farrenheit and radiation similar to a gamma ray burst or extreme x rays. Therefore while the area of the earth, venus and mars are compatible with life the earth is most compatible. Despite this as venus and mars could very well have been incorporated in the same gamma ray burst as the earth extremophiles may very well exist on venus as they did on the earth 3.5 billion years ago. It is also possible that extremophiles may exist under the surface of Mars either in bacterial or viral form.

Our moon which stabilizes the wobble of the 23.5 degree tilt of the earth's axis may have been formed by a planetary collision 4 to 4.5 billion years ago. This wobble if not stabilized would cause a similar phenomenon as with Mars ;evaporation of the atmosphere ;which is why depletion of the ozone layer from excessive solar flare activity could possibly change the angle of tilt of the earth from 23.5 degrees to 22.5 degrees if the mass increases or 24.5 degrees if the mass decreases of the polar ice cap. This may not be sufficient to cancel out the moon's effect on the poles but could still affect the earth.

The moon(our moon) may have been caused by a planetary collision 4 billion years ago. A likely candidate for the collision is Venus which acted like two billiard balls hitting each other on a pool table where Venus was initially a part of the earth(their components at that time were virtually identical) and Venus spun off towards the sun with a trajectory which was altered by angular momentum causing it to orbit the sun in the same plane but in a different direction. Venus's orbit is counter to the orbits of other planet which indicates that it was hurled into that spot of the solar system. Indeed if Jupiter is the sun's binary star rather than a planet; a portion of Jupiter could have been a huge flare like object which impacted the earth causing it to split form Venus(which was thrown off) and the moon which stayed in a stable orbit.If this is true Venus's origins were actually part of the earth and extremophilic actvitiy on Venus is of high probability more than mars.

HOW DOES DARK ENERGY AND THE ACCELERATING EXPANSION OF SPACE-TIME AND GALAXIES OUT WITH 'VACUUM ENERGY' TIE IN WITH THE COSMOLOGIC CONSTANT? CHAPTER SIX BY DR.MITCHELL WICK

Dark Energy was squeezing down to 10^-24 cm in the quantum bubble pre Big Bang and was a major cause of" The Big Bang" however as the Cosmologic Constant is only approximately 10^-55 joules how does that translate into the massive anti-gravitational Dark Energy pushing the galaxies and space-time outward in all directions at an ever increasing rate.? The answer is in "The First Event" of the Time Oscillation Paradox as explained in Megaphysics III;Nothing Doesn't Exist" and The Nth Power" both written by this author.

If one uses M theory one sees that the D-0-branes always existed as an infinite number of completely parallel planes before the first dimension occurred. The force that prevented these d-0-branes or parallel planes from touching each other and forming the first dimension was the Cosmologic Constant or the energy density of a vacuum. While a small number to separate two completely parallel planes from touching and forming the first and then after the time oscillation paradox the 2nd,3rd,4th ... to the nth dimensions in a spiral vortex of the space-time continuum when one realizes that there were an infinite number of completely parallel planes which required 10^-55 joules or the cosmologic constant to separate each plane the energy of the Cosmologic Constant or 10^-55 joules times Rayo's Number approaching infinity (as nothing doesn't exist)causes Dark Energy which approaches infinity times the cosmologic constant as space-time continuously flattens out in all directions with the cosmologic expansion so Dark Energy is in actuality the Cosmologic Constant from pre-first event times infinity=approaching infinite kinetic energy as an asymptote. All the extant dimensions flatten out from an infinite curvature point to asymptotic flatness of flat space-time as the mass contained in expanding space-time approaches the B.M.R. which hets space up by 2.74 degrees kelvin. Th mass forming the cosmologic constant before the First Event was from fermions which were almost massless quantum fields formulating electromagnetic radiation which is also almost massless and the force holding the fabric of space together(when time was infinitelydilated or stopped constricting space.

Infinite parallel planes of D-0-branes from touching each other when anti-gravity pre-dominanted before the Time Oscillation Paradox or the First Event forming the spiral space-time continuum. After the First Event spin2 vector bosons formed from the tachyons that dropped below the speed of light forming the effect of gravity and formulating the –

$\Lambda = R\ ab - \frac{1}{2} R\ ga\ b = -\frac{1}{R2} = \frac{8\pi G}{c4} T\ ab.$ *Prior to the formation of inertial mass* $R\ ab = 0$ *so* $\Lambda = 0 + \frac{1}{2} R\ g\ ab = -\frac{8\pi G}{c4} T\ ab$ *and the* $R\ ab =$ *the inertial mass of space or leptons gluons tachyons and probably photons according to quan*

Quantum field theory which measured a confluence of quantum fields with the potential energy of space and the kinetic energy of the cosmologic constant. The expression 1/R2 is an approximation of space-time curvature as -1/R2 is 0 for an infinite curvature for black holes and R=0 for space-time curvature of deep space's near vacuum as 1/0(squared)=infinity for flat space with R=0 so 0 squared is 0 and 1/0=infinity.RECIPROCAL CURVATURE OF SPACETIME BETWEEN ANTIGRAVITY AND GRAVITY IS FOR THE EXPRESSION 1/R2 for anti-gravity and -1/R2 for gravity as reciprocals you get R2/R2 or

$\infty \div \infty =$ *everything except zero as nothing doesn't exist.*The minus sign in space-time curvature represents RECIPROCAL CURVATURE FOR GRAVITY AND GRAVITY AS COMPARED TO THE +sign for anti-gravity or Dark Energy. This indicates reciprocals as everything except zero equals 1.

HOW THE EINSTEIN FIELD EQUATION WHICH INTRODUCES STRESS ENERGY COMBINED WITH SPACE-TIME ILLUSTRATES THAT THE QUANTUM GROUND STATES IN A RELATIVISTIC UNIVERSE IS THE COSMOLOGIC CONSTANT OR THE ENERGY DENSITY OF A VACUUM AUTHOR; Dr. Mitchell Albert Wick

First of all gravity by definition is the curvature of space-time caused by mass. Gravity is in actuality an effect rather than a force and gravitons are that component of the mass curving space that gives the effect.

Stress energy relates angular momentum in a way that illustrates that as the angular momentum changes so does the relationship of inertial mass and non-inertial mass whereby the inertial mass is the resistance to push and pull by definition and the angular velocity changes as space-time is being curved by the inertial versus the non-inertial mass which causes the effects of gravity and anti-gravity.

In the most famous Einstein Field Equation G ab= R ab -1/2 R g ab for gravity and G ab= R ab+1/2 R g ab for anti-gravity these equal to the Einstein or Gravitation coupling constant"K" times the Stress Energy tensor of that system or systems. This results in $8\pi\ G/c4$ times the Stress Energy tensor which formulates curvature or reciprocal curvature of space-time depending upon whether we are discussing the effect of gravity or anti-gravity which curve space-time inward for gravity and outward for anti-gravity e.g. black holes mass curves space-time inward from flat space to a point of infinite curvature and the near vacuum of deep space curves space-time outward toward asymptotic flatness.

The cosmologic
constant Λ *is the energy density of a vacuum or ϱ vac and is the equation of*

gravity is added to the Ricci Tensor describing inertial mass or R ab – ½ R g ab reflecting the effect of gravity curving space-time inward from no curvature to infinite curvature of a point. The Field Equation is G ab= R ab – ½ R g ab $+\Lambda\ g\ ab =$
$\frac{8\pi G}{c4}$ $T\ ab$ *where the minus sign illustrates that gravity is curving space − time inward toward a point from aymptotic flattness or no curvature with an*

infinite diameter.

The quantum ground state or eigenstate of energy is not 0 but Λ *and as the*

Direction matters $+\Lambda$ *and* $-\Lambda$ *with a magnitude of* 10^{-55} *joules the equation* is
$\Lambda\ g\ ab = R\ ab + \frac{1}{2} R\ g\ ab =$
$-\frac{8\pi G}{c4} T\ ab$ *for anti* –
gravity which describes Dark Energy and the cosmologic constant.

The positive sign is before the anti-gravity notation of ½ R g ab and flatten space-time outwardly towards asymptotic flatness as in deep space being PUSHED outwardly by Dark Energy and is described by the Cosmologic Constant.

For space-time being curved outwardly in all directions G ab= R ab + ½ R g ab-$\Lambda\ ab$

=-$8\frac{\pi G}{c4}$ *T ab indicating recirocal curvature of space – time on stress energy*

as in the vacuum or near vacuum of deep space with near flat space-time and approaching infinite diameter. This illustrates the anti-gravitational effect of Dark Energy as described by the cosmologic constant and DOES NOT EQUAL ZERO which the Einstein Tensor G ab is supposed to illustrate so the ground state energy level is the Cosmologic Constant or the energy density of a vacuum which is not zero.

The effect of gravity curving space-time inwardly as in the event horizon of a black hole is G ab= R ab-1/2 R g ab $+\Lambda\ g\ ab = \frac{8\pi G}{c4} T\ ab = 10 - 55$ *joules not zero*

And also equals 1/R2 whereby R is the curvature metric of space-time. In this case
1/R2=1/∞ 2 $= \frac{1}{\infty} =$
0 *for a* $+\Lambda\ g\ ab$ *in the case of ANTI –*
Gravity and illustrates why the sign of the cosmologic constant must be +

as $+\Lambda = \frac{1}{R2} = 8\pi G(-1)T\ ab$ *and the square root is* $\sqrt{\frac{1}{R2}} = \sqrt{\frac{8\pi G}{c4}}\ (i\ )\ T\ ab$ This is
1/R($2\pi \frac{G}{c2}$ (i) $\sqrt{T\ ab}$ *which mimicks the Action formula of any metric as* -
1/2k2$\sqrt{-g\quad R}$ *where* $K = \frac{8\pi G}{c4}$ *and the minus sign of the metric g is* $i2 = -1/$

Type equation here.

The -
$\Lambda\ g\ ab$ *does not equal +*
$\Lambda\ g\ ab$ *as* $\Lambda\ g\ ab$ *isn'tzero so as the magnitude of the energy density of a vacuum is equal but (*

Opposite the direction is still
$+\Lambda$ *in the anti –*
gravity case which is before the FIRST EVENT and and shows that the quantum ground state
0 *and the kinetic energy of space from the negligible mass of leptons and gluons k parallelpl*

IF LIGHT IS BENT BY MASS OR THE EFFECT OF GRAVITY THEN IS IT TRUE THAT ALL ELECTROMAGNETIC RADIATION IS BENT BY MASS?

As is known space-time is curved by mass and the portion of mass that curves space-time is what is described as the gravitons which cause the effect of gravity. Now as a postulate if all space-time was Mintkowski Space-time asymptotically flat as in the near vacuum of space then it would be electromagnetic radiation or photons which are bent or curved by mass rather than the space which describes the path of the radiation.

Dark Matter described 80 percent of all matter measured in this universe causing the rotation of galaxies to be slower than is projected by Newton's Law of Gravity. Despite this Dark Matter still cannot be directly measured although at this time a measure of mass effects after the Big Bang or Big Bounce and perhaps Inflation shows a clumping and streaking of space in excess of what it was initially after the Big Bang whether or not it was the first event. Dark Matter's effect would curve space-time more than it would be if Dark Matter didn't exist. If this curvature was an effect based on curvature of electromagnetic radiation rather than space-time then Dark Matter would be RIFE WITH ELECTRMAGNETIC RADIATION curving flat space-time far more than with Newtonian Effects however the measurements SHOW THE OPPOSITE. DARK MATTER ILLUSTRATES NO LIGHT OR ANY OTHER KIND OF ELECTROMAGNETIC RADIATION so either electromagnetic radiation doesn't curve flat space and dark matter doesn't exist or electromagnetic radiation does curve space-time because the curvature illustrating the effect of gravity gives a curved path in the presence of mass for which electromagnetic radiation follows and dark matter exists without electromagnetic radiation.Light is bent by gravity as the curvature of spacetime for which it follows is bent by the mass for which it approaches.

DO TIME CRYSTALS MAKE UP 'GOD'?

If the creator comprised of tachyons as a nervous system the Higgs Field as a body and dark matter as a peripheral nervous are indeed immortal only immortal components can produce an immortal product. Time crystals may be the building blocks of the oversoul and indeed all souls.

HOW DID LIFE INITIALLY BEGIN?

BY DR.MITCHELL ALBERT WICK

ABSTRACT

Approximately 3.5 billion years ago the first chromosomes on the planet earth formed which was the blueprint for rudimentary life. Prior to this there was messenger RNA and transfer RNA as well as DNA(desoxyribonucleic acid).DNA,mRNA, and tRNA are inert(non-living) but chromosomes are living. What was the catatlyst that reschuffled the DNA and RNA into chromosome bodies?

3.5 billion years ago the atmosphere around the earth contained methane CH4,hydrogen sulfiteH2S,carbon dioxideC02,oxygen(02),nitrogen(N2) and water (H20).This was not the atmosphere that surrounds the earth in 2017.Also if there was an ozone layer it was significantly smaller than today, The surface temperature of the earth was closer to venus than earth perhaps 350 degrees farrenheit.There was a catalyst. Approximately 3 billion years ago there was a gamma ray burst from the fusion of two neutron stars which cut through this noxious atmosphere like a knife through butter lasting 1-3x10^-3 sec. This gamma ray burst reschuffled the RNA and DNA breaking it into fragments moving them around like a deck of cards being schuffled and in this case there was gravity effects(curvature of spacetime due to mass)which forced an oscillation,spin and resassembly of the DNA into the formation of 46 chromosomes..The sugars from lipids and carbohydrates developed as well as more complex proteins after the first rudimentary life formed.

HOW TO SEARCH FOR LIFE ELSEWHERE

Locate regions of gamma ray burst in the past and present then find celestial bodies which have water,oxygen and carbon. This undertaking may be difficult as finding the locationsof Gamma Ray Bursts of very short duration with our technology may determine locations in the present but finding them from millions of years ago may be very difficult unless a footprint or signature of the gamma ray burst can be found.The size of gamma ray bursts can be very small or as large as a solar system.

HOW DOES DARK ENERGY AND THE ACCELERATING EXPANSION OF SPACE-TIME AND GALAXIES OUT WITH 'VACUUM ENERGY' TIE IN WITH THE COSMOLOGIC CONSTANT? CHAPTER SIX BY DR.MITCHELL WICK

Dark Energy was squeezing down to 10^-24 cm in the quantum bubble pre Big Bang and was a major cause of" The Big Bang" however as the Cosmologic Constant is only approximately 10^-55 joules how does that translate into the massive anti-gravitational Dark Energy pushing the galaxies and space-time outward in all directions at an ever increasing rate.? The answer is in "The First Event" of the Time Oscillation Paradox as explained in Megaphysics III;Nothing Doesn't Exist" and The Nth Power" both written by this author.

If one uses M theory one sees that the D-0-branes always existed as an infinite number of completely parallel planes before the first dimension occurred. The force that prevented these d-0-branes or parallel planes from touching each other and forming the first dimension was the Cosmologic Constant or the energy density of a vacuum. While a small number to separate two completely parallel planes from touching and forming the first and then after the time oscillation paradox the 2nd,3rd,4th ... to the nth dimensions in a spiral vortex of the space-time continuum when one realizes that there were an infinite number of completely parallel planes which required 10^-55 joules or the cosmologic constant to separate each plane the energy of the Cosmologic Constant or 10^-55 joules times Rayo's Number approaching infinity (as nothing doesn't exist)causes Dark Energy which approaches infinity times the cosmologic constant as space-time continuously flattens out in all directions with the cosmologic expansion so Dark Energy is in actuality the Cosmologic Constant from pre-first event times infinity=approaching infinite kinetic energy as an asymptote. All the extant dimensions flatten out from an infinite curvature point to asymptotic flatness of flat space-time as the mass contained in expanding space-time approaches the B.M.R. which hets space up by 2.74 degrees kelvin. Th mass forming the cosmologic constant before the First Event was from fermions which were almost massless quantum fields formulating electromagnetic radiation which is also almost massless and the force holding the fabric of space together(when time was infinitelydilated or stopped constricting space.

How The Left Shift is Affected by the 2.74 degree kelvin Discrepancy caused by the Background Microwave Radiation from "The Big Bang"
By Dr. Mitchell Albert Wick

The Boltzmann Equation measured the entropy(S)of a state and is a fluid transport equation for matter and energy. Space- time is a perfect fluid and energy or photons act as a Perfect Gas. The Boltzmann Equation is S=k ln W where by k=Boltzman Constant or 1.38062x10^-23 joules/degree kelvin and W=number of microstates of the system. This mentions the number of ways molecules of the system in Thermodynamics can be arranged.

To find the wavelength and frequency of electromagnetic radiation at 2.74 degrees kelvin we must multiply 2.74(1.38062x10^-23)in the MKS system to factor label out degrees kelvin from the equation.
E=h
υ *accordingly the energy of a photon equals Plank'sConstant or* $6.63x10^{-34}$*joule* – sec *times the frequency of the electromagnetic radiation. Thermometers appear to be* 2.74 *deg*
colder than they really are2.74(1.38062x10^-23 joules of energy in the MKS system reveal the energy of the system gained or lost by the 2.74 degree kelvin differential in measurement.E=hυ *so* $2.74(1.38062x10^{-23} = 6.63x10^{-34}(frequency\ change)$

hich is 5.706x10^11 hertz or cycles/sec in the MKS system and this is the reciprocal of wavelength
.$\lambda = 1.75438x10^{-2}$*nanomters while frequency is* $5.70572913x10^{11}$*hertz.*

Microwave frequencies are from 300 Megahertz to 300 Gigahertz so 5.70x10^11 hz is 5.70x10^5 Megahertz or 5.70x10^2 Gigahertz for the background microwave radiation from The Big Bang.Radiowaves are a longer wavelength than micrwaves going from 0.04 inches or 1 mm to 100 kilometers so the corrected frequency of 5.70x10^2 Gigahertz is the upper limit of the frequency of corrected BMR as a boundary to radiowaves which falls also in the area of UHF or ultrahigh frequency which is why "snow "appears as the BMR on a t.v.screen. As a result the distances of galaxies and stars are farther away from us than expected and the expansion of this universe is faster than projected by the uncorrected BMR.

GLOSSARY

Abelian :equations having a coefficient or variety in a specific group,g,,algebraic number fields,tensors of the same degree or cohominy group

Anisotropic :not isotropic,lacking observational symmetyry

Anti-symmetric:tensors or vectors that are equal but opposite and can therefore partially cancel or cancel

Aymptotic:that which approaches a level or degree but never reaches it;asymptotic flatness appears without curvature but doesn't reach it

Bianchi's Identity: The identity of groups of Riemannian 4 space that is anti-symmetric and Abelian and cancels each other out of being equal but opposite
"The Big Bang"A theory proposed describing a Friendman type I open expanding f;at universe with is homogeneous and isotropic

"The Big Swirl "A Big Bang with a progressively decreasing rotational vector from an infinite curvature point of space-time to asymptotic flattness

Black Hole :collapsed matter from a neutron star or galaxy with extreme curvature of space-time at the central nexus due to extreme gravity of of the spiral space-time

Calabi Yau Manifold:a surface which represents a relative isotropic portion of space-time with a puckering to accommodate multiple dimensions considered a twisted variant of the orbifold
Choas:absolute disorder

Chiral:a mirror image or absolute symmetry

Closed string:a two or one dimensional building block of matter from energywith movements in 10 or 26 dimensions without breaking the string

Compactified:when every point of the dimensions are curled up mathematically making the size approach zero.First determined by Kaluza and Klein

Conformal Space:when every point in space relative to every other point maintains its relative position regardless of what the space is doing

Dark Matter;an indirectly measured mass causing perturbations in gravity(the curvature of space-time)caused by mass.Acts as cosmic glue containing possibly baryonic particles and neutrinos

Event Horizon:area where a black hole is perceived by measurementsEntropy:degree of disorder

Entropy:degree of disorder
Ex nihilo:out of nothing

M(Membrane)theory:the 5 dual string theories into one massive theory of everything which incorporates membranes which vibrate and incorporate all energy and matter

Isoropic:observational symmetry

Geodesic:a unit of space-time
Gravity:the curvature of space-time caused by mass;actually an effect not a force

Membranes:a description of matter in terms of energy states with stress energy densities described in the number of states with regard to dimensions

N;number of dimensions in N dimensional space

Open string:a two or one dimensional bulding block of matter with movements in a multidimensional plane

Orbifold:space-time manifold in an open twisted cone configuration utilized in string theory

Relativity:the behavior of matter and energy with regard to other matter and energy;energy and space have a different vantage point from other matter and energy including stress energy,time and mass with changes regarding relative velocity

RicciTensor:that tensor which represents inertial mass or resistance against pull or push
Riemann Forces:all strong and weak forces in nature

Riemannian Space:Mintowski space with Riemann curvature of space-time caused by mass.Flat space if no mass is present

Scalar:the magnitude compone t of a vector or tensor with regard to direction

Six dimensional string manifold:curled up closed strings in configuration according to Kaluza and Klein which is 10-33 cm and may be Calabi Yau Manifolds

HOW DOES TIME WORK IN RELATION TO SPACE?

The following conclusions may be intuitively obvious but could be difficult to visualize so as the author I will attempt to devise a system which incorporate time, space and the rate of sequencing of events.

Time by definition is the sequencing of events. The rate of the sequencing determines if time dilates or is constricted relative to space. If time dilates it constricts space and if time constricts it dilates space. The dilation of space causes a flattening effect on space as in a vacuum and the constriction of space causes what approaches infinite curvature and can constrict space to a point with dilated time curving the space towards infinity. As space-time acts as a perfect fluid it would appear as a progressively decreasing dimple in a whirlpool in an ocean of space-time. In the near vacuum of deep space time is constricted and curvature approaches flatness(no curvature).Constricted time goes infinitely fast and dilated time goes infinitely slow so as a vacuum is approached time speeds up and and one propels into the future. The rate of the sequencing of events increases as time constricts and decreases as time dilates according to the observer time traveler. As time constricts or dilates space the rate of sequencing of events constricts or dilates space. Reversing times arrow reversing the sequencing of events and therefore must reverse the rate of dilation or constricting of space by time(which can be though of as constricting or dilating space like a rope rapping around a ball or a cow).

Going backwards in time can be thought of as traveling to another universe whose first event occurs after the first event in our universe;but as the destination location may not be the same as the origin location the time traveler could land in the vacuum of intergalactic space at a temperature of just above absolute zero 5 billion years in the past relative to the time sequencing of our universe. This can occur by utilizing a wormhole or Einstein Padowsky Rosen Bridge near the event horizon of any active black hole or merging black holes such as the mini black holes in the HadronCollider in Cern,Switzerland.

ONTOLOGIC PROOF by Dr. Mitchell Albert Wick

Visualize a vortex with 2 disparate component fields in solenoids. The fields can be called
ω *and* ω'*. If the field described by* ω' *has a significant gradient to* ω *and* ω *has a chosen*

Volume whereby
ω *is outside and* ω' *is inside there is a significant gradient for the harmonic function*

Between the inner and outer volume. For all
ω *and* ω' *satisfied if at the boundary of regions(R) the derivative of the function*

The Faber-Jackson Relation =normal projection of the vorticies. Using the Tensor Virial Theorme and the vortex bulge systems with the Faber-Jackson Relation
Re^$\alpha \propto$
$\sigma(0)^2$ *which is the region in topologic space of the exponential function of* σ*(space)*

2 dimensions are zero leaving an infinite number of parallel planes whereby
MH$\rightarrow \sigma$ *and* $0 \rightarrow M$.

H=Hamiltonian Operator $=\hbar \frac{}{2m} \quad \nabla$ ^$n \quad n \rightarrow 0 \quad m = mass$
$\nabla =$ *LaPlacean Operator*

MH$\propto \sigma\ 0^4$ *whereby* $4 = \#dimensions.$

The anti-symmetric tensor field of spiral spacetime interacts with the massive Higgs Field and massless Permanent semi-radius tachyons via the LaGrange Identity and LeGendre Transformation using the action formula for strings interacting with the Massive Higgs Field. The Abelian Higgs Model for Abelian Higgs Vorticies is anti-symmetric equal but opposite cancelling via the Gauss-Bonnett Identities such that

$$\Gamma.\nabla\, x)\Gamma = \nabla\, x.\Gamma)\Gamma = -(\Gamma.\Delta\, x)\Gamma \text{ where} \nabla = \text{harmonic function}$$

$$\Gamma = \text{spherical boundary index}$$

The boundaries of harmonic tensor fields relate to spacetime vorticies and the Higgs Field. Anti-symmetric or skew symmetric tensors incolved in Riemannian Spacetime A^t=-A and a ij=-a ji via Bianchi Identittity for spacetime from I to j.For all$\forall$ i to j

i=initial event and j the final event

The leGrange Identity is g det(g I j)> 0.

Spacetime coordinates ds/dt)^2 as in the line element has ds=ds/dt . g I j dx^ i dy j=IType equation here.

I^ j≤ i

The manifolds or surfaces of spacetime can translate into tensor fields so ds^2= g i j. dx ^ i dy ^ j

And (x+y)y'+ x −y =0

g i j = r i r j

g= I j μ^i μ ^ j

The gradient volume inside the Harmonic Function as nabla approaches 0 or the Cosmologic Constant $\nabla x \rightarrow 0$ so $\Gamma.\nabla x)\Gamma \rightarrow 0$ and as $antisymmetric$ $tensor$ $fields$ $(\Gamma.\nabla x)\Gamma \rightarrow 0$ or Λ so $-(\Gamma.\nabla x)\Gamma \rightarrow 0 = \Lambda$ .If - $(\Gamma.\nabla x)\Gamma = -g$ ji and $(\Gamma.\nabla x) = g$ i j

In an anti-symmetric abelian tensor filed g ij = g ji

G ij=r1r2 and g=det(g ij) then (g ij)^2= g i^2 j^2- g(i j) where g I j = volume of the vortex and as (vortex $\Gamma.\nabla x)\Gamma \rightarrow 0 = \Lambda$ and $-(\Gamma\nabla x)\Gamma \rightarrow$ 0 any $vortex$ $including$ $space - time$ of a $harmonic$ $function$ of ∇ x $where$ $\Gamma =$ $index$ of the $spherical$ $boundary$ or $interface$ as 0 $volume$ is $approached$ $equaling$ the $cosmolog$

CHAPTER THREE :TACHYONS AND FERMIONS WITH REGARD TO THE FIRST EVENT;EVERYTHING IS SOMETHING AND NOTHING DOESN'T EXIST

The products of all Fock Spaces for all Harmonic Oscillators
$\ldots \nabla n \ldots q$ *in the tensor virial theorem for the Ontologic Proof of God*)(infinite number) yields an eigen-state
$\Pi n, \mu\{a\, n^{\ddagger}, \mu\}|0>$
whereby the harmonic oscillators yields the ground state or eigenstate $=$
$0 \frac{or\Pi(n=0)1}{2^n} + 1\pi\varpi\, i \rightarrow \frac{j}{2^{2\pi\omega}j} \rightarrow$
ior the infinite number of $D - 0 -$
Branes orthe infinite parallel planes.The spectrum of the lowest lying eigenstates have tach

(infinite parallel planes with infinitely dilated time).Tachyons existed in the lowest eigen-state or the ground state(zero dimensional state)and were an integral part of the space continuum existing at v>c while fermions(the fermionic vacuum state)were at v<c where c=3x10^8 meters/sec.Each and every Fock Space is composed of the energy equivalent of hadrons which are massless as space is massless .The Fermions which are massless comprise of leptons for the fabric of space or the sum total of all Fock Space(down to and even below the Hilbert Space limit which relates with Planck Length(10^-33cm).The quantum dot principle described in this author's first book "Mega-physics :A New Look at the Universe "subdivided below the Hilbert Space limit and as the components of Hilbert and Fock Space is comprised of leptons which are 99.999% potential energy being held in the fabric of space by massless gluons as it's energy equivalent .Of course the ground state energy levels of leptons and gluons as miniscule as they are both virtually massless as also indicating the initial vacuum state prior to the First Event.

Regee Trajectories for open strings illustrate that the tachyon is the God Particle prior to the formation of the spin 2 vector boson after the first event (time oscillation)which formed the Higgs Field .Fock Space for each any every Harmonic Oscillator(;see tensor virial theorem in the Vortex Model Ontologic Proof).
The Fermionic Vertex limit with very small Fock Space(s) is Q BRST|R>=0 with an infinite number of Fermionic verticies .The infinite number of Fermionic verticies→
utilize the same onshell matrix element. Φ *is the vanishing anticommutation relation and BRS*

IS THE CHARGE.V=[Q
BRST,
Φ] *WHERE Q is nilpotent.There are an infinite number of vertex functions in vacuua*

And each vertex function represents a plane of space without time where all planes are totally parallel with a Regge Slope of 0.Theses vacuaa are $e^{\wedge}q\phi(0)|0 \geq |q>$

and the conformal weight is $\frac{1}{2}\varepsilon q(q +$
$Q)$*where the fermionic and vertex finctions are each infinity corresponding to the infinite nu*

ber of parallel planes before the time oscillation paradox between fermions and tachyons which both were part of a continuum forming the spin 2 vector bosons in the vortex of space-time caused by the oscillation of time causing the spin and centrifuge effect forming the string dimensions .V 3/2=[Q BRSTε *V* 1/2] ... *V* 5/
2[*Q BRST*, ε *V* 3/
2] ..... *ad infinitum. All these verticies are equivalent to each other AND TH THEY ARE INFINIT*

FERMIONIC -FERMIONIC SCATTERING AMPLITUDES USING FORMALISM NET MIRROR IMAGE EFFECTS NETTING ZERO.AS THE FERMIONIC STATE IS THE VACUUM STATE WITH AN INFINITE NUMBER OF PLANES ARE NO DIMENSIONS THE FERMIONIC VERTICIES ALL NET THE FUEL FOR EVERYTHING WITH THE TACHYON BEING THE POTENTIATOR WHEN A ROGUE TACHYON DROPS TO THE SPEED OF LIGHT CAUSING THE TIME OSCILLATION BETWEEN TIME's ARROW GOING FORWARD AND BACKWARD.

BASED ON THESE CONCLUSIONS SPACE ITSELF IS COMPRISED OF FERMIONS AND TACHYONS.UNDER THE SPEED OF LIGHT FERMIONS ARE THE AETHER OR QUINTESSANCE AND THE INFINITE NUMBER OF PARALLEL PLANES ARE COMPRISED OF FERMIONS UNDER THE SPEED OF LIGHT AND TACHYONS ABOVE THE SPEED OF LIGHT. It's like a totally blank blackboard with no contents or points(intersection of three planes).The blackboard in this case is comprised of fermions and tachyons with the speed of light boundary being a dichotomy or boundary or interface between tachyons and fermions where the sum total of fermions and tachyons equals the dimension of time once the speed of light boundary was breeched by either fermions or tachyons. Space under the speed of light is comprised of fermions and above the speed of light comprised of tachyons and at the speed of light space-time constricts toward zero as mentioned in my books "Mega-physics II ; An Explanation of Nature" and "The Equation of Everything" .So prior to the first event there we no points as there were no plane intersections and the intersection of three planes form a point .The planes were composed of fermions and tachyons; depending on the velocity and time's arrow(Tachyons backwards ;Fermions time forwards)Space=fermions(tachyons)and are dependent on velocity which form the dimension of time when the fermions and tachyons are mixed .When this occurs the spin 2 vector bosons are formed with mass, the Higgs Field, and time wraps around space constricting or dilating it. Space()no dimensions=velocity of fermions with respect to the speed of light+ or –velocity of tachyons with respect to the speed of light.$\{\ \} = \frac{d\{0\}fermions}{dt} + or - d(0)tachyons/dt$ and c=0 space-time. So c=d(fermions/dt)+ or – d(tachyons)/dt .So as fermions and tachyons comprise space and are the components of space and as the intersections of these special planes are zero in the zero dimensional state; the fermionic space and tachyonic

space were separated by the boundary of the speed of light which has zero space-time .As the fermionic state is the vacuum state and space is comprised of fermions the fermions are leptons and gluons where leptons and gluons are massless as the previous book "What is the Dimension of Time? "proved that space-time is massless, then before the first event the derivative of the sum total of fermions with respect to time=0infinitely dilated) and the derivative of the sum total of tachyons with respect to time=0(infinitely dilated)also equals 0.(note quarks and anti-quarks have a variable mass but quarks and anti-quarks didn't form from gluons until after the Oscillating Time Paradox).Based on the concept that the advent of the time oscillation or paradox occurred at exactly 3x10^8 meters/sec or the speed of light; THIS POINT IS WHERE TIME BEGAN AND WHERE TIME BEGAN CONSTRICTING AND DILATING SPACE. From the Lorenzian Transformation time dilates toward infinity at the speed of light or in essence stops as space-time constricts toward zero at this point. IN THAT MATTER OF SPEAKING TIME'S ORIGIN WAS AT THE SPEED OF LIGHT AS THAT IS WHERE SEQUENCING OF EVENTS STARTED AND IN A SENSE THE SPEED OF LIGHT BOUNDARY COULD BE CONSIDERED TIME WITH SPACE EIXSTING ABOVE AND BELOW THIS BOUNDARY .ALSO AS THE SPEED OF LIGHT IS A CONSTANT NOT A VARIABLE TIME CAN BE ELIMINATED TOTALLY FROM THE EXPRESSION ds/dt as
dt$\rightarrow \infty$ *and* $s \rightarrow \infty$ *as space* $-$ *time constricts toward* 0 *making distance* $\rightarrow$
∞. *THIS IS SYNONYMOUS WITH THE WORM HOLE PHENOMENON*.Note the Harmonic Oscillator in the tensor virial theorem
is
∇x *and* Γ *is the spherical index boundary.* ∇x *reveals the infinite harmonic oscillators involved*

In the Fermionic vertex model and tachyon model revealing the 0 eigen-state or ground state.

Hadrons are composed of bosons and fermions where the Higgs Boson is massive and subcomponents of fermions comprise leptons and the six flavors of quarks(which have mass while gluons are massless)

PRIOR TO THE FIRST EVENT LEPTONS AND GLUONS WERE AN INTEGRAL PART OF SPACE.AFTER THE FIRST EVENT QUARKS, GLUONS AND LEPTONS WERE INVOLVED WITH SPACE AFTER THE TIME OSCILLATION WHICH FORMED THE VORTEX OF SPACE-TIME WITH THE STRING DIMENSIONS.SPIN 2 VECTOR BOSONS ALSO FORMED AFTER TIME CONSTRICTED AND DILATED SPACE.TACHYONS INCORPORATE WITH THE HIGGS BOSON IN THE 'GOD PARTICLE' AFTER THE FIRST EVENT AND EXISTED WITHOUT THE HIGGS FIELD BEFORE THE FIRST EVENT.

Quarks are composed of strings as they have mass. Strings also have mass so the next question is did strings always exist or have their inception only after the first event? Spin 2 vector bosons only existed after the first event which described gravity which is the curvature of space-time caused by mass. Mass must exist as a pre-requisite for gravity to exist. Therefore as quarks have mass, they curve space-time which means it must be a post-requisite of the first event. As strings have mass albeit negligible ;they must post-requisite the first event ;as are photons which have a negligible mass which isn't zero .Also the formation of the string dimensions must form at the same time as strings .However space must be composed of something that pre-requisites strings without it being nothing but must have zero gravity. This therefore can't be quarks, mesons or even neutrinos; although baryons may qualify .Despite this GLUONS FORM QUARKS AND ENCOMPASS A SUPERSTRONG FORCE AND THE FRACTURE OF GLUONS CAN INVOLVE A SINGULARITY LIKE A BIG CRUNCH OR BIG RIP WITH ENORMOUS FORCE FAR GREATER THAN THE STRONG FORCE. GLUONS ARE MASSLESS AS ARE LEPTONS AND THE GLUONS MAY HOLD SPACE-TIME TOGETHER WHILE THE LEPTONS ARE THE PARITITONS OF SPACE-TIME WHICH ARE BEING HELD TOGETHER.THE FABRIC OF FLUID SPACE IS COMPRISED OF LEPTONS AND GLUONS HOLD SPACE TOGETHER.AS THE LAW OF CONSERVATION OF ENERGY STATES THAT ENERGY CAN NEITHER BE CREATED NOR DESTROYED THE COMPONENTS OF SPACE SUCH AS GLUONS, LEPTONS AND TACHYONS MUST BE COMPRISED OF ENERGY AS THEY ARE MASSLESS.ENERGY ALWAYS EXISTED AND ALWAYS WILL EXIST. THE GLUONS AND LEPTONS WERE INITIALLY 100% POTENTIAL ENERGY HOLDING TOGETHER THE FABRIC OF SPACE AND COMPRISING THE FABRIC OF SPACE UNTIL THE FABRIC OF SPACE WAS DISRUPTED AND TRANSFORMED BY THE TIME PARADOX CAUSED BY A ROGUE TACHYON TRAVELING AT V=C WHERE THE PARADOX OCCURRED.

The distance between each and every manifold in the infinite plane hypothesis approaches zero without reaching it as it is the zero dimensional state. Basically, the static energy of the components of space comprised of leptons and gluons acts as a "cushion" between the planes or manifolds to prevent the manifolds from touching each other(.See diagram 1).It also keeps the manifolds intact.

$R\infty(a,0,0,0) =$
$\textit{the Region in Topological space whereby the potential energy of gluons holding the fabric of}$

Space together is "a" being acted upon by b,c,d where b,c, and d are the tensors which are zero(0).R a refers to the potential energy of the sumtotal of all the manifolds or planes which are cushioned by the potential energy of each and every other plane or manifold.Therefore R(a,0,0,0)=-R(a,0,0,0) and the region of topological space of R+(-
R)→
$0.\textit{If the sum was zero it would define dimensions yet it must be zero as the scintilla between}$

each manifold would be space-less-ness and as energy always existed and requires space to exist;this option isn't possible; therefore the partition between each manifold must be a membrane which adheres to each manifold or surface by the potential energy of the gluons. So as a result D-0-branes=R(a,0,0,0)where a is the gluon potential energy adhering the manifold to the D-0-brane and preventing the manifolds from being in contact with each other which is a requirement for being perfectly parallel and therefore the zero dimensional state.

Note there may be a negligible repulsive force between the energies of leptons and gluons acting as the cushion;however this would only occur if there was a flattening effect of space as time was infinitely dilated resulting in anti-gravity as suggested by the Cosmologic Constant or 10^-53 or the energy density of a vacuum or the vacuum state rather than the false vacuum of deep space therefore there must be a repulsive force suggestive of anti-gravity even in the absence of any measurable mass from leptons or gluons.$R\infty(a,0,0,0) or\ the\ Region\ of\ Space = \Lambda\, a + \Sigma\, D - 0 - BRANES\ whereby\ the\ potential\ energy\ of\ the\ D - 0 - Branes = potential\ energy\ of\ leptons + potential\ energy\ of\ gluons \rightarrow 10^{77} joules\ as\ mentioned\ in\ this\ author's\ previous\ book$ What is the Dimension of Time. The cushion between each and every parallel plane or manifold is caused by the composite of the D-0-Brane composed of the potential energy of the gluons and leptons and the anti-gravity effect of the Cosmologic Constant.Total$\rho = \rho\ vacuum + \rho\ (leptons) + (gluons). \rho\ vacuum = \Lambda\ and\ \rho(leptons + gluons) = 10^{77} joules\ as\ 100\%\ potential\ energy\ which\ converted\ to\ kinetic\ energy\ after\ the\ time$ Paradox which formed the space-time vortex and string dimensions .If the leptons and gluons were 100% kinetic energy it would either be contained in the infinite parallel planes or would leech into the region of topologic space between the infinite parallel planes . Considering all the possibilities there is sufficient kinetic energy to leech into the area between the parallel planes which are repelled by the anti-gravity space flattening effect of the Cosmologic Constant. As this area approaches zero without it being zero this kinetic energy component would be very small call it $\in (K.E.) \rightarrow 0\ which\ separates\ each\ and\ every\ D - 0 - Branes.$Mathematically therefore
$\Lambda\ and\ \rho(leptons + gluons) = 10^{-53} + \ \rho(99.9999 - \epsilon)P.E. for\ leptons\ and\ gluons) + \rho(\epsilon)K.E. for\ leptons\ and\ gluons.$Remember that the Law of Conservation of Energy states that energy cannot be created or destroyed making energy a continuum(without beginning or ending) and as infinitely dilated time still exists as time and time always existed time is part of the space-time continuum. As a consequence time and energy always existed as well as space and they will always exist.

Utilizing E=mc^2 as an approximation ;at isn't exact .....if E=10^77 joules and c^2=3x10^8 m/sec)^2 then the mass=10^77/9x10^16=1.1x10^93 kg of mass in all the universes in the space-time continuum .As a consequence of this as the mass of this universe is 10^64Kg then 10^29kg is spread out through the rest of the space-time continuum.Despite this, there is empirical evidence of approximately 10^54kg as the mass of this universe leaving 1.1x10^103 kg mass in the space-time continuum.As 10^103 kg is inclusive of dark matter based on space-time curvature the other 10^29 kg is the forward momentum of space-time being acted upon by the momentum of other universes in the multi-verse. Again as the correction factor for hadrons and photons is m^2c^4+mc^2 then the mass becomes

10^77/c^6=10^77/3x10^48=1.3X10^125 kg.)^1/2 which is over a googlplex of mass as the components of mass of composed of hadrons. Because of this there are approximately 1.22x10^125kg)^1/2 in our universe/10^54kg in our universe universes based on mass. The number of universes in the multiverse was discussed as universes in the space-time continuum based on the correction factor of Einstein's equation as applied to massless or near massless hadrons traveling at or near the speed of light .Despite this, the mass of our universe is the square root of 1.3x10^125 kg or 1.1x10^63Kg which has extreme precision from the known value of 10^54 kg utilized in this author's book "Mega-physics II:An Explanation of Nature" and "What is the Dimension of Time? "Physicists at Stanford University Andre Lindy and Vitalie Valchurin mentioned the extreme possibility that there are 10^10^16 and maybe the possibility of 10^10^10^7 or e ^10 e^10^10^7 which are far greater than a googolplex. Based on calculations however, there may be 10^10 universes in the multi-verse based on differential mass calculations.On other interesting point relates to the radius of this universe as being approximately 10^10 light-years.Does this mean that the 10^10 universes in the multi-verse have an average radius of 1 light-year;it seems more likely that the velocity of light changes throughout different formations of space-time curvature between the universes of the multi-verse twisting and turning based on the varied masses and being bent as gravity(curvature of space-time caused by mass)making it appear as though the mean radius of other universes is one light year when in reality the mass density of these disparate universes is greater and at times far greater than than our own which is primarily a near vacuum state. As an example;in older universes there may be a good deal more black holes than in our universe and this would skew the reading of space-time into a more spiral configuration making the radius and diameter(Schwarzchild)appear smaller than it actually is.This is more evidence of Obler's Paradox as proposed by Edgar Alan Poe who said that our universe is very young as the night sky is mostly dark rather than full of starsIt also implies that the Second Event was not necessarily the simultaneous formation of all the universes but had s sequential order in which our universe was later in the sequence.As a result the relative position of our universe in the space-time continuum while still close to the center of the vortex which is pulling us toward it at an accelerated rate may have other universes closer to the center of this huge space-time vortex formed from the First Event.Much of this will be discussed later.. This does not totally rule out the possibility of the existence of one universe however provides soft conclusive evidence that 10^64 kg reveals the total mass of all universes as space-time is relatively massless indicating that the frequency of black holes may either diminish or evaporate at a greater rate in other universes and that the formation of stars and galaxies may be in an earlier stage in some other local universes.It also indicates that a Big Crunch indicated in the book "What is the Dimension of Time?" accelerating the expansion geometrically by pulling the galaxies with a velocity of 2.2x10^35(c)meters/sec with the push of Dark Energy if forcing homogeneity and RELATIVE ISOTROPISM RATHER THAN ABSOLUTE ISOTROPISM with reference to the observer such as the WOMP or Hubble Space Telescope and that the Big Crunch is pulling the mass apart on a string level to fill the void of accelerating space-time in a diffusion like process as space-time is a perfect fluid. This means the mass is like a particle in a suspension in motion and as space-time increases in velocity homogeneity will increase ,black hole evaporation may increase and the discernable edge will blur out more and more .This will make it possibly more difficult to determine the supermassive black holes including the site of "The Big Bang" unless space-time curvature measures are used as in the LIGO PROJECT.THE HAWKING PARADOX ABOUT INFORMATION BEING LOST IN AN EVAPORATING BLACK OUT IS ANSWERED AS THE DIFFUSION OF THE INFORMATION AT THE EVENT HORIZON OF BLACK HOLES INTO FLUID LIKE SPACE-TIME IS LIKE THE DISSOLUTION OF COFFEE INTO WATER.THE INFORMATION BECOMES A UNIFORM SUSPENSION IN FLUID SPACE-TIME AS WITH HAWKING RADIATION AND DIFFUSES OUT IN ALL DIRECTIONS AROUND THE POINT OF ORIGIN OF THE DIFFUSION WHICH IS THE EVENT HORIZON AND THE SPUMING OUTWARD OF THE QUASAR.THIS MEANS THE INFORMATION FROM EACH AND EVERY EVAPORATING BLACK HOLE IS SMEARED THROUGHOUT SPACE WITH THE CONCENTRATION OF THE SMEAR DECREASING GEOMETRICALLY FROM THE POINT OF ORIGIN OF THE SMEAR ;information is preserved only extraordinarily difficult to detect.This is due to the "traffic jam"effect of matter existing in dilated time relative to the observer at the event horizon however over several hundred thousand years the information will "bleed through"enough to harmoniously diffuse out into space-time from the point off maximum constriction at the event horizon(Schwarzchild Space-time)to space which appears to be more and more asymptotically flat as the effect of the flattening of the Cosmologic Constant takes over.

CHAPTER FOUR :THE HARMONIC OSCILLATOR AND THE TIME PARADOX

The Kalb Raymond Action is the most likely pathway for any metric describing an open or s=closed string.As an action it has the highest probability magnitude for the path integral of the function of the tensors of the first,second third and fourth degree going up to the nth degree as described by the open set which contains the subsets from a1 to n where
$n <$
∞ *regarding eigenstates of energy over n dimensions. The statistically probability distributio*

Follows the quantum mechanical distribution of any event defined by the metric "g"which is measurable.Heuristic algorhythms follow probability distributions of the likelihood of each and every event with regard to space-time and therefore can easily be used in Quantum Mechanics although "fuzzy logic"can be used for the case of AI or artificial intelligence. Statistics is important as well as q values which involve error measurements or describe regions of error due to measurement error made by the intelligent observer. If the intelligent observer is part of what's being measured the results will be skewed and if the Higgs Field is what's being measured the intelligent observer must in the case of "The God Field"must be the tachyon or permanent semi-radius tachyon which acts as the "brain"of the Higgs Field and is separated from the Higgs Boson by the boundary of the speed of light or "c". Using "fuzzy"logic associated with Heuristic Algorhythms shows a distribution curve with the height of the curve or the maximum height being the Kalb-Raymond Action for the closed or open string;the the tensor virial theorem is applied to apply to the vortex of the Higgs Field which incorporates the vortex of the space-time continuum. Heuritstic algorhythms apply to the permanent semi-radius tachyon as the brain and the thought component of the Higgs Field is based on choices and outcomes based on these choices which have a perfect bell shaped probability distribution. AS THE ACTION FORMULA FOR ANY METRIC 'g"can follow a bell shaped probability distribution which is NOT RANDOM(space-time curvature metric is related directly to the –g or minus metric to the half power)then the action formula follows heuristic algorhythms.The effect of creation from the Time Oscillation Paradox was caused by a "rogue"permanent semi-radius tachyon dropping to the spped of light from over light speed.Was this a chaotic action or was it deterministic. It was deterministic if there was conscious thought impelling the permanent semi-radius tachyon to drop to the speed of light. Were their choices or was this a random event? Based on Heuristic Algorhythms there is a set of probabilities for tachyons to act in a random manner and another set of tachyons which acted out of choice.{permanent semi-radius tachyons}={permanent semiradius tachyons[random],permanent semi-radius tachyons[deterministic or by choice]}.IF ONLY ONE ROGUE TACHYON DROPPED TO THE SPEED OF LIGHT FROM ABOVE LIGHT SPEED THE PROBABILITY OF THE ACTION BEING DETERMINISTIC IS GREATER THAN BEING RANDOM OR CHAOTIC.IF MANY PERMANENT SEMI-RADIUS TACHYONS DROPPED TO THE SPEED OF LIGHT AS A GROUP OR SUB-GROUP WHICH WERE INTERRECONNECTED THE PROBABILITY OF RANDOMNESS IS INCREASED.PLEASE NOTE THE SECOND LAW OF THERMODYNAMICS WHICH STATES THAT ENTROPY FOLLOWS WHERE THINGS GO FROM A MORE ORDERED

STATE TO A LESS ORDERED STATE OR CHAOTIC STATE.AS GROUPS OR SUBGROUPS OF PERMANENT SEMI-RADIUS TACHYONS ARE LESS ORDERED OR HAVE A GREATER ENTROPY THAN A SINGLE TACHYON THE PROBABILITY OF THIS BEING A CHAOTIC ACTION IS INCREASED.There is no clear way to establish which of these quantum states are correct especially if the groupings of permanent semi-radius tachyons are interconnected as in a brain,however this configuration would indicate THOUGHT ACTING ON THE HIGGS FIELD WHICH WOULD INDICATE ORDERED CHAOS(Mega-physics;A New Look at the Universe).The confluence of tachyons can act like cell assembies(Hebb's Theory of Neurobiotaxis)which act similar to neurons or a neural net with logic subsystems. In this case what may appear to be random or chaotic may in actuality be deterministic.

THE MASS OF ANTI-PARTICLES AND THEIR PROPERTIES

This article is a detailed description of the mathematical exposition describing the properties of antimatter including a variable mass .It will also delve into practical applications of matter -antimatter systems as a potential energy source and the risks involved with an unsecured system.

Antiparticles have been experimentally determined in CERN , Switzerland to have a positive mass .According to Newton's Law of Gravitation Force of gravity=G m1m2/r 2 or the force of gravity equals the gravitational constant G times the product of two masses divided by the distance between the masses squared. To get a negative gravitational force or antigravity would require one of the two masses be a negative mass .As matter and antimatter annihilate that would cause a modest repulsion between matter and anti-matter .As matter and antimatter attract it is unlikely that antimatter has a negative mass. However, using math the Force of gravity can also be negative if the mass of two particles is as such mass/i mass2/i so the radius squared becomes a -1(radius) squared. A mass for antiparticles could be and has been mathematically proven as the mass of ordinary matter/i or the square root of -1,This mass is a hybrid between positive mass and negative mass and oscillates during and throughout the distance from the positive non- oscillating mass. This oscillation of the mass coupled with the collision with positive mass triggers the annihilation reaction between antiparticles and particle yet it would still register as positive mass when weighed but would qualify as strange matter having the strange property of oscillating hybrid mass between negative and positive.

BACKGROUND

This author is able to mathematically prove using the Harmonic Oscillator that antimatter has a variable mass which is why after annihilation with matter strange matter in the form of quarks and gluons result. Particles and antiparticles have the same quantum number but opposite charge and quantum spin. Matter anti-matter annihilation result in the production of gamma rays ,force carrier particles such as gluons and a W/Z force carrier particle. Positron and electron annihilation result in gamma rays with a rest energy of .511 Mega electron volts1 ;however protons and antiprotons result in mesons which can decay into neutrinos ,positrons electrons and more gamma rays.

Physicists at the Brookhaven Laboratory have determined the existence and detected strange quarks in antimatter nuclei. The Relativistic Heavy Ion Collider(RHIC) detected an anti hyper-triton 2 which has an antiproton ,antineutron and anti-lambda particle .The anti lambda particle is an extremely heavy anti-nucleus and is produced when gold ions are collided at very high energies producing quarks ,antiquarks ,and gluons which when cooled produce hyperons ,up quarks and down quarks .These subatomic particles are all classified as hadrons .Particles produced by these collisions produce particles below the N-Z plane produced by nuclei ;N being the number of protons ,N the number of neutrons and S the degree of strangeness of the subatomic particle .Evaluation of strange quarks which are or=proposed to be in the collapsed neutron stars which are pre-black hole can determine what happened at or about Planck Time 10-43 seconds at 'The Big Bang "when a plasma soup of quarks was formed including up-quarks ,down-quarks ,and strange quarks. **MATHEMATICAL EXPOSITION**

Mathematically the Harmonic Oscillator can fluctuate between 0 degrees or 0 radians and 180 degrees
or
π *radians where the sine of these extremes is* 0 *and the sine of the extremes of* π *and* 2π *radia*

Is also 0 but from a mirror image deflection of the energy state being measured or operated on with regard to time .The Hamiltonian operator$\mathcal{H}$=p 2/2m+kx 2 and the Energy of the system =(n+1/2)ω

whereω *is the angular velocity* $n =$ *the quantum level of the system* $p =$ *momentum indicator which* $=$ *mass times velocity* $k =$ *the spring constant and* x *is the period displacement.*

In an oscillating antimatter mass p=1/2(mω)1/2(r+r')where r has canonal commutation$[p, x]$ and the Hamiltonian operator$\mathcal{H}$=1/2ω(rr'+r'r).For positive mass p=1/2(m$\sqrt{m\omega}\ (r + r')cos\theta\ where\ \theta =$ $o\ degrees\ or\ o\ radians. For\ negative\ mass\ .p =$ $\sqrt{2m\omega\ \ (r + r')cos\theta\ where\ \theta = \pi\ radians}$ or $\sqrt{2\ \ m\omega}$(r+r')cos θ where $\theta =$ $\pi\ radians$

P=momentum

$\omega =$
$angular\ velocity\ and\ the\ x\ coordinate\ along\ with\ x'will\ be\ the\ complex\ conjugate\ of\ p\ amd\ p'.$

X=i(2mω)$\frac{1}{2}-(r-r')cos\theta\ for\ positive\ mass\ x =$

$i(2m\omega) -$

$1/2\cos\theta\ and\ for\ negative\ mass\ x' = i(2m\omega) - \frac{1}{2} - (r-r')cos\theta - where\theta$=o

radians for positive mass and $\pi\ radians\ for\ negative\ mass.\cos 0 = 1\ cos\pi =$

$-1\ so\ it\ follows\ that\ p = \frac{\frac{1}{2(m\omega)1}}{2(r+r')times}1\ or\ p = \frac{m\omega 2}{2(r+r')1}and\ p' =$

$\frac{m\omega 2}{2(r+r')} - 1. The\ complex\ conjugates\ x\ and\ x'are\ i(2m\omega) - 1/2\cos o(r -$

$r')and\ x' = i(2m\omega\} - \frac{1}{2}cos\pi(r-r')$ -1/2=1/square root of mass x angular

velocity+or-1 (r-r')I NOTE ;cos $\theta\ is\ in\ the\ numerator\ for\ o \leq \theta \leq \pi\ and\ r =$

$radius\ of\ rotation$

That is if

2mω) $to\ the\ plus\ and\ minus\ one\ half\ power\ respectively\ or\ their\ sqaure\ roots.$

The zero point energy equation 3 is the Hamiltonian2 Operator$\mathcal{H} = \int_0^{\pi} d\sigma(PuXu) -$
$L\ where\ Xu\ has..above.$L=p uX u-1/2 $e(p2u + m$ 2as the first order form of the Hamiltonian operator at ther zero point energy.$\sigma relates\ to\ the\ equations\ of\ motion + \tau\ to -$
$\tau\ where\ \tau\ relates\ to\ relative\ time.$
Also ,using Poisson's Equation

∇2=4$\pi\rho\ for\ a\ dual\ vector\ field\ (r,r')oscillating\ between$;=4$\pi\rho cos\theta$= 4$\pi\rho$(+1)

and 4$\pi\rho$(-1) or 4$\pi\rho$ and $4\pi - \rho\ in\ the\ dual\ vector\ field\ revealing + and -$

$energy\ density\ of\ mass.$The dual vector field is (r ,r') for =ρ and (r' ,r)for –ρ.In the zero point energy phase or using the massless null vector, an Abelian anti - symmetric tensor field can give a super-space formulation using tensor gauge theory with a massless Abelian anti -symmetric tensor field of rank 2 .The stress energy tensor T ab can in oscillating mass(which nets out as zero mass between =1 and -1)can be shown as anti commutative as T ab=T ba nd T ab=-T ba such that T ba=-T ba ,–T ab=T ba,-T ab=T ba and T ba=-T baas T ab cos0=T ab T ab cos π=-T ab T ab cos π=T ba,T ba cosπ =-T ab.In the zero point energy state the stress energy tensors are Abelian and anti -commutative.3

CONCLUSION

The momenta and coordinates of the oscillating mass represented by the complex conjugates$[p, x]$ of the oscillating mass nets zero in the zero point energy stage.

PRACTICAL APPLICATIONS:

At .511 mega electron- volts from collisions matter anti-matter annihilation could possibly be a superior energy source to nuclear power if and only if containment of gamma rays will render it safe. The cost of such traps and screen may also be prohibitive and while plants may be strategically located in thinly populated areas and the energy transmitted to quadrants of various countries can an accident be avoided? In medicine positrons are used now in nuclear imaging of metastatic sites called PET scans , Positron Emission Tomography which help determine whether or not treatments for metastatic carcinomas are effective or not. The mechanism of action of the PET scan involves injection of the isotope Flu-deoxy-glucose(18

F)which is also known as FDG which is chemically mixed with radioactive isotope fluorine-18 substituted for a hydroxyl group in glucose as a positron –emitting radioactive isotope. A PET scanner can detect FDG in its distribution throughout the body. The 18F-FDG is taken up by the brain ,kidney and malignant cells where phosphorylation stops glucose release preventing further metabolism so the molecule undergoes further radioactive decay while in the cell. Finally the 2'-flourine group is converted to 18 O-;which is heavy oxygen, which can be excreted normally.4As fuel the price of converting antimatter into high energy matter would be prohibitive albeit extremely efficient if enough antiparticles could be gleaned for use as a fuel.

G.R. Schmidt, H .P. Gerrish and J .J. Martin at the NASA Marshall Space Flight Center in Huntsville ,,Alabama have proposed Antimatter production for propulsion applications ..They propose anti-protons as a catalyst in fission based thrust production. The energy cost of 1 microgram can be affordable at 6 million dollars and can be used for spaceflights and intra-stellar space missions based on antimatter catalyzed fusion.5.This article by the above delves into six alternative antimatter catalytic propulsion applications.

APPENDIX

The weak, strong and null energy conditions indicate that a positive mass rather than negative occurs in nature .Using path integrals with a flow-line integral over curve C for the null vector and using the stress energy tensor T ab the path integral

$\int c = T\ ab\ kakb\ d\ \lambda \geqq$

0 *where a and b are superscripts of k and k is the null vector.. The weak energy condition stat* that for each time a vector field is observed non-negative density of matter is

perceived. The above uses Poisson's Equation which states that the derivative

operator of a dual vector field equals

$4\pi\rho$ *wherep is the energy density of matter.* $\rho = T\ ab\ x\ n\ xb \geqq$

0 *andn and b are superscripts to x. Every future point vector field –*

T b a y b where a and b are superscripts in the tensor equation must be future point with rega

To times arrow, The strong energy condition has every future pointed vector field
measures non-negative. T ab-1/2T g ab X a X
b≥
o. again a and b are superscripts too x. Note: The Inflation Theory of Alan Guth and a scalar f
Field with a positive potential violate this.7 The Casimir Effect also violates the

negative mass exclusion by matter .The Casimir Effect shows that$-\rho = \epsilon =$

$-\frac{\pi 2}{720}\frac{h}{d}4$ *where* $h =$ *Planck'sconstant* $\epsilon =$

energy density of matter and d 4 is 4 dimensional spacetime. 6

FOOTNOTES AND BIBLIOGRAPHY

1.Wikipedia.Annihilation,http :en. wikipedia.org/wiki/Particle Annihilation 1.1 p.2

2.PHYSICS WORLD.COM RHIC Nets Strange Antimatter March 5,2010.p.1 and 2
3.A Superspace Formulation of Abelian Anti-symmetric tensor gauge theory
.Modern Physics Letters A 15(2000)p.965-978
4..Wikipedia. Fludooxyglucose(18F)Properties and Mechnism of Action. .http: en
.wikipedia.org/wiki/Fludooxyglucose
5.Antimatter Production for Near-term Propulsion Applications .Smith and Mayer
Penn sylvania State University.
6 Wikipedia Energy condition.http:en.wikipedia.org/wiki/Energy condition u
7.Guth,Alan .The Inflationary Universe.www.Edge.org
1.Kaku,Michio. Introduction to Superstrings and M Theory.1.8 p.37-38 Harmonic
Oscillators Springer Press 1998.
2.ibid.,see footnote 1

3.ibid see footnote 2

4.ibid see footnote 3

5.ibid see footnote 4.

6.ibid see footnote 5.

7.ibid see footnote 6

8.ibid see footnote 7

GLOSSARY AND DEFINITIONS

Abelian :an abelian group also known as a commutative group is the result of a group operation of 2 group elections which doesn't depend on their order.i.e., a set (a, b)combine to form another element a . b.()=operationof elements a and b.

Anti-symmetric or Anti-commutative:-
a= $a \quad \leftrightarrow \quad a =$
$0. Swapping\ the\ position\ of\ 2\ arguments\ or\ elements\ negate\ the\ result. \sigma = group\ \forall \alpha \epsilon \sigma$

Casimir effect:a small attractive force that acts between two parallel uncharged conducting plates or physical forces arising from a quantized field. Described as zero point energy of a quantized field in the intervening space between the objects

The PROPERTIES OF ELECTROMAGNETIC RADIATION AS MATTER

BY Dr. Mitchell Albert Wick

ABSTRACT

According to Albert Einstein light is bent by gravity. This can only ne possible if photons have mass. It has been scientifically determined that the mass of a photon is not zero albeit miniscule. 1 At or about absolute zero photons have been slowed down in the laboratory to 36 miles per hour 2;this deceleration cannot occur unless photons have a measurable mass. Because photons have a measurable mass they can be considered matter as the equivalent mass from the de Broglie Equation$\lambda = h\frac{}{mc}$ where λ *is the deBroglie Wavelength*.. Based on this equation and using the Boltzmann Equation one can prove that electromagnetic radiation occupies the gaseous state of matter and the plasma state of matter depending on the environmental temperature. Also one can demonstrate that in black holes radiation occupies all states of matter including liquid, solid, and Boso-Einsteinian condensate.
Once shown that radiation is gaseous one can apply Boyle's Law and Charles' Law eo illustrate that as a perfect gas it follows the properties of any gas at standard temperatures and pressure (STP). Then one can apply the Perfect Fluid Equation for space-time with radiation as a perfect gas to attempt to predict radiation behavior in different media at super high pressures or at temperature extremes.
vrbo.com

BACKGROUND

THE BOLTZMANN EQUATION *describes a fluid transport equation not in equilibrium according to statistical mechanics. The entropy of a given state S=k LnW where k=Boltzman constant1.38062 x10 -23 joules/degrees kelvin and W is the number of microstates of the system. It represents the number of ways molecules in a thermodynamic system can be arranged.4.Considering the Perfect Fluid Equation for space-time manifolds and photons of electromagnetic radiation in the form of a PERFECT GAS,ONE CAN USE AVOGARDO'S NUMBER 6.023X10 23 MOLECULES/MOLE OF AN IDEAL GAS AT STANDARD TEMPERATURE AND PRESSURE in the statistical mechanics of the Boltzmann Equation over six dimensional space-time to determine the entropy of a macro-state involving an ideal gas(electromagnetic radiation)in the microstates of the system W and then compare it with Steven Hawking's figure for black hole entropy of 0.29 in which 252 separate states of matter exist according to S=N(*$\sqrt{\ }$ $Q1Q5$ where N is the number of states Q1 and Q 5 are the* *charge differentials of the states in a black hole or Q1=one brane and Q5 is a 5 brane.N should approximately equal W in the Boltzmann Equation as the number of microstates of the macro system. Statistical thermodynamics reveals that the number of microstates corresponding to each photon is W where W=N!/∏iNi! WHERE Ni= the microscopic condition of position and momentum (p). Given electromagnetic radiation as an ideal gas of N photons the probability of each state is equal to ||W||2 or*

W=N!/∏iNi! $The\ force\ and\ diffusion\ theorem\ the\ derivative\ of\ f =$
$total\ change\ of\ collisions\ according\ to\ the\ Lionville\ Theorem\ and\ Hamilton's equations\ of\ colli$

Df/dt where d is the partial derivative of the total change with regard to time with respect to collisions$\Delta td33rd3p = \Delta fd3rd3p\ where\ F(r + \frac{p}{m}t, p + f\Delta t\ and\ t +$
$\Delta tare\ the\ derivatives\ of\ N\ the\ number\ of\ collisions\ equaling\ \Delta fd3rd3p\ where\ d\ 3\ is\ three\ dime$

Total phase space. THE COLLISION LOSS AS A FORCE FIELD ACTING ON PHOTONS IN SPACETIME AS A PERFECT FLUID+F(R,T) WHERE M=MASS OF THE PHOTONS OR 10-23 grams giving the partial derivative of f with regard to time + p/m$\Delta f + F\ the\ partial\ derivative\ of\ f\ with\ regard\ to\ p(momentum) =$
$the\ partial\ derivative\ of\ f\ with\ regard\ to\ t\ with\ respect\ to\ collisions\ and\ if\ no\ collisions\ occur$
Between photons you get the Liouville Equation which is the collision loss of photons in phase space acting on t where t=time.

THE PERFECDT FLUID EQUATION IS THE RATIO OF PRESSURE (P) TO ITS ENERGY
DENSITY$\rho\ WHERE\ W =$
$\frac{p}{\rho}. POISSONS\ EQUATION\ STATES\ THAT\ THE\ DERIVATIVE\ OPERATOR\ OF\ A\ DUAL\ VECTOR\ FIEL$
$4\pi\rho\ AND\ THE\ COMBINED\ GAS\ CONSTANT\ IS\ .0821\ WHERE\ \rho MRT = \rho Mc2. C =$
$the\ thermal\ speed\ of\ molecules = \sqrt{\ } RT\ where\ T =$
$degrees\ kelvin\ for\ a\ perfect\ gasand\ w = \frac{p}{\rho} = \frac{pC2}{pc2}\ where\ C = c\ therefore\ as \frac{C}{c} =$
$1\ w =$
$\frac{p}{\rho} for\ a\ cold\ gas\ of\ electromagnetic\ radiation\ or\ the \frac{pressure}{t} heenergy\ density\ of\ matter\ or$
radiation of matter is unity for W in other word
$\rho =$
$p\ the\ energy\ density\ of\ matter\ equals\ the\ pressureof\ an\ ideal\ gas. The\ energy\ density\ of$
matter of electromagnetic radiation as determined by
frequency(
$v\ equal\ its\ pressure\ \ the\ phase\ space\ with\ regard\ to\ time\ of\ spacetime\ in\ the\ perfect$

fluid equation Q.E.D.

Laser Technology involves a matrix whereby photons are trapped in a gaseous ,liquid or solid matrix . In a sense lasers are light amplification in a gaseous or liquid suspension and can be used as a catalyst in trapping as little as a pictogram of photons in an environment similar to the super pressures of a black hole as a new energy source or fuel .Laser stands for light amplification by stimulated emission of radiation. Optical amplification through stimulated emission of electromagnetic radiation is how a laser emits light columnated overlarge distances with coherence..A gain medium is excited by a source and a gain medium adsorbs energy to be trapped between two resonators. A photon of radiation repeats in the gain medium and is emitted through an aperture.

Boltzmann's H Theorem

H is the probability integral over what is known as velocity space. H is the quantitiy of events or metrics fpr N statistically independent particles. With v=velocity in d3 space

H=$\int \quad P(\ln P)d3\ v =< ln\ p >$

$or\ the\ expectation\ value\ of\ the\ natural \log of\ the\ probability\ of\ N\ particles\ moving\ at\ velocity$

$S = entropy = -NkH\ where\ k =$

$the\ Boltzmann\ constant. For\ electromagnetic\ radiation\ v =$

$c\ and\ P\ approaches\ but\ doesn't reach\ \ 1. Therefore\ as \ln 1 = 0\ H =$

$\int \quad p(\ln P)d3v = 0\ As\ s = -NkH\ and\ H =$

$0\ the\ entropy\ of\ this\ system\ for\ a\ black\ hole\ approaches\ zero. So\ according\ to\ the\ H\ theorem$

N particles moving at v=c in a black hole or the quantum bubble pre ‘Big Bang"has a statistical probability of 100%for interaction of events or metrics with approaching zero entropy. In a black hole entropy is postulated by Steven Hawking as 0.29 and approaches the statistical conditions reached by the Boltzmann H Theorem. Motion between photons is dampened to the degree that a liquid solid and Boso-Einsteinian condensate phase of matter can be reached.

BOUNDARIES AND INFINITY by Dr.Mitchell Wick

WHEN ONE LOOKS AT A CIRCLE specifically it's circumference there is no beginning or ending. " The Equation of Everything" compactitifies to a circle with the circumference being space-time the radius being positive Riemann Forces of Nature radiating to the center from the circumference and negative Riemann Forces of nature from the center to the periphery or circumference.. Space-time is asymptotic to s spiral or cone as a confluence of circles going from a point of infinite curvature to flat space-time with an infinite diameter. Of course the diameter of the point approaches zero and the diameter with flat space-time approaches infinity all without reaching them as asymptotic relations. When these circles are placed adjacient to each other from a point to near infinite diameter flat space they form half a cone with the mirror half being passed a boundary making a bi-conal raltionship which is asymptotic to a spiral with the rotation being counterclockwise in one region of space-time and clockwise in it's mirror image. Stephen Hawking stated that "there are no boundaries or that the only boundary is that there are no boundares"and in this author's first book it was stated that "the only boundary is that of spacelessness".Basically these axioms are both stating the same thing as a boundary to everything is nothing and nothing doesn't exist as it is spaceless. It has been stated that any event or metric cannot be infinite as it "makes the math blow up into nonsense"yet the circumference of a circle(not a sphere)is infinite as it has no starting point or endpoint.Pi to the last digit is infinite so to use pi in an equation would be incroprating infinity and yet pi is an integral part of everything as is the Cosmologic Constant. Rayo's number is the largest non-infinite number and can be utilized as the upper limit of energy with the cosmologic constant as the lowest number and Hilbert Space or Fock Space can be considered a lowest unit of space-time with regard to string theory relating to the Orbifold or below 10^-33 cm. Of course the orbifold rotates with mirror directions forming an open spiral with one pint forming a sphere approaching a point of infinite curvature while the outer open end forms curved space-time utilizable in the action formula of Einstein. The big question regards how entanglement affects eigenstates of energy acting upon or being acted upon by Riemannian or Lorenzian Space-time as a finite function rather than infinite. When anything is finite it must be bounded and if it's bounded then what's outside the boundary might have "nothing"or spacelessness or might be of a different character with different rules. Spacelessness is impossible which is that which is beyond the comprehension of mankind and "The Equation of Everything"has the Cosmologic Constant as the ground state energy level with the uppoer limit being that of the HiggsField or 10^77joules or the Grand Unification Energy 10^19 giga electron volts or Rayo's number as infinite energy seems improbable although there is Potential Energy associated with space as near massless leptons and gluons appear to form the fabric of space and kinetic energy equal to the Cosmologic Constant or 10^-55 joules when prevent an infinite number of completely parallel planes from touching each other to form the first then second and third dimensions during the first event as the spiral space-time continuum is formed. According to the Law of Conservation of Energy nothing doesn't exist as the potential energy of space and the kinetic energy of the Cosmologic Constant get converted into other forms of energy during and after the Time Oscillation Paradox of the First Event. If entanglement of different eigenstates of energy is finite then the combinations or permutations of these disparate energy states is finite and therefore bounded. This entanglement function must be incorporated into each energy state between the action of each string as a separate metric in each region of space-time as reflected as an orbifold in Hilbert or Fock Space. As without entanglement "The Equation of Everything "states there are 524,288 equations in nature these equations when self-entangled are still not infinite but based on the measured effect by some measurer relating to each event which must relate to each action for every metric(measured value requiring a measurer).Also the presence of the measurer affects the measured value as is true in photon pairing and Spooky Action at a Distance. Still the result must be infinite to be analogous to the circumference of a circle for compactification of the equation to be a circle and as space-time must be approaching infinity(as nothing doesn't exist) the circumference which is infinite is approached by space-time which approaches being infinite and can be described as a tensor to the fourth degree over a finite number of eigenstate of energy going from the ground state or Cosmologic Constant to Rayo's Number which incormporates the Higgs' Field energy of 10^77 joules with the GrandUnification Energy of 10^19 Gev and then has self entanglement with each disparate energy state which incorporates each an event event as an action of each and every string to string interaction in both deSitter and anti-deSitter Space,Fock Space and may actually be smaller than the Hilber t Space limit. Still is there a clear lower boundary to space and string to string interactions? Is there are membrane which encompasses a lower boundary to string-string interactions? How does the function of entanglement affect string to string interactions in both a Heterotic 8x8 SO-32 type I II and IIA string theories forming Mtheory relating to membranes.Are all of the entangled strings twisting and turning Fock ,deSitter and anti-de-Sitter space formulatin all the permutations and combinations relating to entanglement and is it still finite?

BIBLIOGRAPHY

1:Peat,F.David. Superstrings and the Search for the Theory of Everything.Yang Mills Forces p.114

2.Kaku,Michio. Strings,Conformal Fields and M theory.Ising Model p.176-78
3 : Wald ,Robert .General Relativity.Chicago,Ill. University of Chicago Press.1984 4.CPT THEOREM; Quantum Field Theory Kaku ,Michio
4:Metric tensor(General Relativity)Wikipediaa and Spacetime.en.m.wikipedia.org.spiral Space-time Einstein 1912 Fractal Time.p.108-109 Braden ,Gregg 2009 Library of Congress HAWKING RADIATION. Wikipedia
5.Peat,F.David .Superstirngs and the Search for the Theory of Everything.p.106-107.Calabi Yau Manifolds

6.Kaku,Michio .Quantum Field Theory. Renormalization Actions in Quantum Field Theory
7.Peat,F.David .Superstrings and the Search for the Theory of Everything.
8.Kay, David C. Tensor Calculusp.129 Osculating Plane

9.Kaku, Michio. Strings,Conformal Fields and M theory.

10.Peebles,P.J.E .Principles of Physical Cosmology.p

11Chang,Alan. HAMILTON JACOBI EQUATIONS UNIVERSITY OF CHICAGO 2013.Zeno's paradox: The Math Forum at Drexel University

12:Tipler,Frank j.The Physics of Immortality

13.Godel,Kurt.Godel's Incompleteness Theorems en.m.wikipedia.org

16:ibid item#7 p.237-42

14.Randall,Lisa. Warped Passages
15:Green,BrianThe Elegant Universe.
and
14.Wick,Mitchell Albert .Megaphysics ,A New Look at the Universe.

15:Kay,David.CTensor Calculus.
18:Kaku,Michio. Strings, Conformal Fields and M Theory.
19:Wikipedia.Electronen.wikipedia.org/wiki/Electron
20:Spooky Action at a Distance Quantum Entanglement Wikipedia. Or en.wikipedia.org/wiki/Quantum entanglement
21:Hau,Len. Harvard Research circa 2003.

BIBLIOGRAPHY

Barrero,John D.The Anthropic Cosmological Principle.Oxford England.Oxford Press.1986
Brade,Gregg.fractal Time 2009 Library of Congress.
Greene,Brian.The Elegant Universe.NewYork.Vintage Books editor Random Press.1999
Hawking,Steven and Penrose,Roger.The Nature of Space and TimePrinceton,N.J;Princeton Science Library 1996
Kaku,Michio.Quantum Field Theory.A Modern Introduction.Oxford university Press.1993
Kaku,Michio.Strings,Conformal Fiels,and M theory 2nd edition.Springer Press.2000.
Kay,David C.Tensor Calulus Schaum's Outline Series. N.Y.McGraw Hill 1998.
Peat,F.David.Superstrings and the Search for the Theory of Everything.Chicago.Contemporary Books 1998
Peebles,P.J.E.Principles of Physical Cosmology.Princeton Series in Physics.Princeton University Press 1993
Wald,Robert m.General Relativity.Chicago,Illinose.University of Chicago Press 1984
Wikipedia: on lin encyclopedia.
Randall, .Lisa .Warped Passages HarperCollins Publishers.N.Y.2005
Tipler ,Frank J. Physics and Immortality. Anchor Books division of Random House.1993

16:ibid item#7 p.237-42

16:ibid item#7 p.237-42

S=rθ the equation of arc length when applied to the osculating plane will pptro an infinite number of dimensions.The unit tangent vector of a sphere of

2

π*radians in motion is* $2\,\pi\,R\cos\theta$ *where* θ *is* π *radians.The sphere travels at HOR THE HUBBLE* EXPANSION FACTOR.As θ *approaches zero* $\cos 0$ *appraches* 1 *causing space* − *time to approach* $-\frac{1}{2e} - i$ *times the number of dimensions where* $e =$ 2.71828 *then spacetime* =

$-\frac{1}{2}e -$

i(to the n power)x1 where n approach $\frac{es\infty\frac{or1}{-1}}{2ei\infty}$ *power or* 0 *which indicates that n infinite number*

Down to theta approaches zero causes r or the arc length at the circumference to approach zero

CONTINUATION OF MATHEMATICAL EXPOSITION

kl

R ijkl - R ji=-8πG e (ij,ji) =g

ji

-8$\pi Ge(ij,ji)$=g ji

vector product of Rij 'Rji=e(ij,ji) cos π =-e(ij,ji)

$8\pi G = \Lambda$ -1 is from cos θ $\theta = \pi$ klR ij.R ji ji for space-time curvature from antiparticle of metric g ji on antiparticles Rji kl=contra-variant tensor kl on covariant tensor ji ji is from antiparticle g ji kl is from gravity of contra-variant tensor for metric g ij for matter

Ijkl=antigravity effect on particles from antiparticles

2

R ij Rji 2 2

--. e ji where Rij Rji/$\left\|R \quad ij\right\| \left\|Rji \quad\right\|$=cos θ

2

$\left\|Rij \quad\right\| \quad \left\|Rji \quad\right\|$

2

add e kl to e ij e kl= R kl /$\left\|R\ kl \quad R\ kl\right\|$ ' g kl cosθ ' g kl=-1 . g kl where$\theta =$ π *radians as covariant tensor For contravariant tensor* $e\ kl = R\ kl$ *squared*/ $\left\|R\ kl \quad R\ kl \ .g - g\ kl = g\ lk\right\|$

kl

R ijkl = g kl/g ij'gkl=-

8$\frac{\pi G}{-16\pi G} = \frac{1}{8\pi G}$*whioch is the reciprocal of the cosmologic constant* Λ $8\pi G =$

$6.67x10\frac{11}{8\pi} =$

$7.5x10$ 10 *joules* –

seconds which is a huge antigravity effect of Dark Energy from the antimatter antimatter e

The gravitational coupling constant k=-$8\pi G/c4$ -$8\pi G/c$ 4 is also synonymous with antigravity between 2 particles of matter or anti-matter .As matter- matter interactions have a positive gravity anti-matter anti-matter interactions are antigravity Q.E,D.

Using E=mc2 with the number of strings in the quantum bubble for antimatter and matter being the mass the calculation emerges as E=7.5x 10 10 joules(3x10 10 cm/sec)2=6.75x10 31 joule-sec or on one second 6.75 x10 31 joules as the blast force from "The Big Bang" ;as antigravity is the predominant blast force Dark Energy blasted out in the 360 degree Orb Blast producing 750 billion galaxies from the Mas string calculation. Note 10 31 is 10 to the thirty-first power

NOTE: Although some sources list the cosmologic constant (∧) as $8\pi G/c^4$ this has no impact on the mathematical calculations only labeling. Other sources list this as the gravitational coupling constant.
G is in kg-meters/sec 2

6.67x10 11/8π =7.5 x 10 10 for number of strings. Calculation or at 10 7 erg per joule for 'Big Bang' blast force is e=mc 2 or e(joules)=7.5x10 10(3 x 10 10)2=75 x 10 30 joules or at 10 7 erg per joule 75 x 10 37 erg or 7.5 x 10 38 erg

www.ingramcontent.com/pod-product-compliance
Ingram Content Group UK Ltd.
Pitfield, Milton Keynes, MK11 3LW, UK
UKHW022015190726
13853UKWH00005B/1949

9 798507 739400